D1616998

Northeast Lakeview College
33784 0001 2332 3

Chemical Process Research

Dedication

This book is dedicated to the memory of Max Rey of Cilag AG, Schaffhausen, Switzerland and the University of Zurich.

He was a dear friend who added so much to our lives, an outstanding chemist who added so much to process chemistry, and a great teacher who added so much to the future generations.

Ahmed F. Abdel-Magid

ACS SYMPOSIUM SERIES **870**

Chemical Process Research

The Art of Practical Organic Synthesis

Ahmed F. Abdel-Magid, Editor
Johnson and Johnson
Pharmaceutical Research and Development, LLC

John A. Ragan, Editor
Pfizer Global Research and Development

Sponsored by the
ACS Divisions of Organic Chemistry and
Medicinal Chemistry

American Chemical Society, Washington, DC

Library of Congress Cataloging-in-Publication Data

Chemical process research : the art of practical organic synthesis / Ahmed F. Abdel-Magid, editor, John A. Ragan, editor.

p. cm.—(ACS symposium series ; 870)

Includes bibliographical references and index.

ISBN-13:978-0-8412-3824-4

ISBN 0–8412–3824–3

1. Pharmaceutical chemistry—Congresses.

I. Abdel-Magid, Ahmed F., 1947- II. Ragan, John A., 1962- III. Series.

RS401.C46 2003
615′.19—dc22 2003060061

The paper used in this publication meets the minimum requirements of American National Standard for Information Sciences—Permanence of Paper for Printed Library Materials, ANSI Z39.48–1984.

Distributed by Oxford University Press

PRINTED IN THE UNITED STATES OF AMERICA

Foreword

The ACS Symposium Series was first published in 1974 to provide a mechanism for publishing symposia quickly in book form. The purpose of the series is to publish timely, comprehensive books developed from ACS sponsored symposia based on current scientific research. Occasionally, books are developed from symposia sponsored by other organizations when the topic is of keen interest to the chemistry audience.

Before agreeing to publish a book, the proposed table of contents is reviewed for appropriate and comprehensive coverage and for interest to the audience. Some papers may be excluded to better focus the book; others may be added to provide comprehensiveness. When appropriate, overview or introductory chapters are added. Drafts of chapters are peer-reviewed prior to final acceptance or rejection, and manuscripts are prepared in camera-ready format.

As a rule, only original research papers and original review papers are included in the volumes. Verbatim reproductions of previously published papers are not accepted.

ACS Books Department

Contents

Indexes

Preface

The development of a drug is a long and complex process that requires a high degree of teamwork and collaboration among multiple disciplines in a pharmaceutical company. The lead process in drug development is chemical process research, which is responsible for identifying the best route to prepare the needed amounts of a chosen drug candidate for all other operations (e.g., toxicology, formulation development, and clinical studies). This research has to be accomplished in a timely fashion to meet the development needs and deadlines. A synthesis procedure has to be developed that will deliver the drug candidate safely, economically, and in high yield; additionally, the synthetic procedure must consistently meet preset physical form and analytical specifications. The development of a suitable procedure starts with an existing method, usually the synthesis used by medicinal chemists in their initial preparation of the drug candidate. This synthesis may or may not be appropriate for large-scale production. In some cases, the existing method may be completely inappropriate, for example; if the synthesis is prepared by solid phase methods, or if the compound utilizes prohibitively expensive or toxic reagents. In this case, a new synthesis is designed "from scratch" to meet the above requirements. In other cases, the synthesis may be partially adopted, with changes to a few steps to obtain a suitable process. Finally, in some cases, the synthesis may remain as is, with only minor changes in reaction conditions and work-up procedures. In all cases, elimination of any hazardous conditions, solvents, and/or reagents is necessary to convert the synthesis procedure into a process that can be scaled safely.

It is important to emphasize that the need for synthetic optimizations at this stage is not in any sense a negative reflection on the chemistry developed by our medicinal chemistry colleagues; their goals are well-served by the original synthesis. Their syntheses are designed to prepare small quantities of products and to cover a wide range of structural diversity. It is not advantageous for the medicinal chemist to spend several months seeking a more efficient synthesis of a particular compound that is just one of many potential drug candidates that may or may not proceed further in development. It is more productive to focus their efforts on preparing a wide variety of related structures to move the program forward as quickly as possible. Once a lead compound is identified as a potential candidate and moved into early development, process chemists can then address the issues associated with its synthesis.

The goal of chemical process research in general may be described as the transformation of a synthesis procedure capable of producing, at best, grams or even hundreds of grams of product at high cost to a highly efficient process capable of safely producing much larger quantities at a lower cost. Chemical process research is the advanced science that often gives the impression of being an art. It transforms syntheses from inoperable to practical, from hazardous to safe, and from expensive to economical. The end result usually captures chemists' attention and admiration and demonstrates clearly the "art" of putting the vast chemistry literature to practical use. Most importantly, the end result means making health care and quality of life affordable to more patients.

The symposium upon which this book is based, *The Role of Organic Synthesis in Early Clinical Drug Development,* was held at the 222nd national meeting of the American Chemical Society (ACS) in Chicago, Illinois in August 2001. This on-going symposium is unique because it is sponsored by two of the largest ACS divisions: Medicinal Chemistry and Organic Chemistry. This cosponsorship is a realization of the increased interest in the role of chemical process research in drug development. The symposium provided a forum to present some of the recent advances in chemical process research by top researchers in the field representing different pharmaceutical companies. This book is meant to gather several of the recent advances in chemical process research into one volume. The ACS Symposium Series, which is designed for rapid publication in areas that are changing and developing rapidly, is ideal for this format. It allows the expansion of the scope of the symposium to cover additional subjects of interest.

This book contains 11 chapters on different aspects of chemical process research written by lead researchers in the field. The authors present examples from several pharmaceutical companies to show the different styles of chemical development research. Yet their goal remains the same: to prepare the desired products safely, economically, and in a timely fashion. The opening chapter in this book is of particular interest to chemists everywhere. It is written by Edward J. J. Grabowski, Merck Research Laboratories, one of the most noted and admired process chemists. The chapter presents reflections of the author on his long and illustrious career at Merck, his thoughts on an ever-changing industry, and an overview of several drug development processes. The remaining chapters describe specific and unique processes that demonstrate the abilities of process chemists to accomplish their tasks.

The goal of producing this book is to be helpful to synthetic organic chemists in industry and academic institutions as well as graduate students. It provides a reference for current topics in chemical process research and may be used as an educational tool. The different styles of chemical development, the variety of structures, and the real-world application of different techniques should appeal to professionals in process research and to organic chemists in general.

Acknowledgments

The idea of organizing a process chemistry symposium came from the collaboration of the ACS Divisions of Organic Chemistry and Medicinal Chemistry. We thank the two divisions for sponsoring the symposium. Our particular thanks to Cynthia A. Maryanoff (Johnson & Johnson PRD), and Lisa McElwee-White (University of Florida), Takushi Kaneko (Pfizer Global Research and Development) and William Green-lee (Schering Plough Research Institute) for their guidance and help during the organization of the symposium. We also thank Keith DeVries (Eli Lilly) for helping with the organization.

We thank each of the authors whose outstanding contributions to this book made it such a worthy project and a valuable addition to the literature. We appreciate their professional courtesy, patience, and kindness, which made this an enjoyable experience.

We are especially thankful to those who contributed both as speakers in the symposium as well as authors in the book: Vittorio Farina of Boehringer-Ingelheim, Todd S. McDermott of Abbott Laboratories, Margaret Faul of Eli Lilly & Company, Jayaram K. Srirangam of Pfizer-LaJolla, and Jerry A. Murry of Merck Research Labs. We thank those who contributed as speakers in the symposium: Kirk Sorgi of Johnson & Johnson PRD, Martin Karpf of Hoffmann-La Roche, Switzerland, and

Edward Delani of Bristol-Myers Squibb. We also thank those who contributed chapters to the book: Edward J. J. Grabowski of Merck Research Labs for writing the opening chapter, John E. Mills of Johnson & Johnson PRD, David Damon of Pfizer Global Research and Development, and Bert Willemsens of Janssen Pharmaceutica, Belgium.

We appreciate the financial support to the symposium by the following companies: Johnson & Johnson PRD, Pfizer Global Research and Development, and Wyeth Research.

Last, but not least, we thank the ACS Books Department staff, Bob Hauserman, Stacy VanDerWall, and Kelly Dennis in acquisitions and Margaret Brown in editing and production for their help and outstanding efforts in producing this book.

Ahmed F. Abdel-Magid
Drug Evaluation–Chemical Development
Johnson & Johnson Pharmaceutical Research and Development, L.L.C.
Welch and McKean Roads
Spring House, PA 19477
amagid@prdus.jnj.com

John A. Ragan
Chemical Research and Development
Pfizer Global Research and Development
Groton, CT 06340
john.a.ragan@groton.pfizer.com

Chapter 1

Reflections on Process Research

Edward J. J. Grabowski

Merck Research Laboratories, Merck & Company, Inc., Rahway, NJ 07065

#Dedicated to my early Merck Mentors – Ed Tristram, Roger Tull, Pete Polack, Erwin Schoenewaldt, Mike Sltezinger, Earl Chamberlin, John Chemerda, Vic Grenda, Seemon Pines and Len Weinstock – a truly formidable bunch of teachers and scientists.

The achievements of an organization are the results of the combined efforts of each individual. **Vince Lombardi**

The kind invitation of the Editors of this Symposium Series to provide an introductory chapter affords me an excellent opportunity to reflect on what has been a thirty eight year career in Process Research – all at the Merck Research Laboratories. There have been monumental changes in organic chemistry during this time, and today the field of Process Research is being pulled in more directions than ever before. Regardless, my belief is that our principal objectives remain the same: the design of practical chemical processes to support drug development, and the discovery and definition of new synthetic methods to support process activities.

When I was interviewing for a job in 1965 as the completion of my doctoral research approached, my thought was to be a medicinal chemist and engage in that alluring area called 'basic research'. I suffered all of the preconceived notions about 'process research', and thought that I would have none of it. A number of the process groups that I visited clearly fit the stereotype, until, of course, my interview with the Process Group at Merck. Max Tishler organized the group along non-typical lines, and there was a very strong emphasis on developing new science, studying reaction mechanisms, and participating in the greater chemical community. These activities were a natural part of the group's principal responsibility – the design and development of superb chemical syntheses and manufacturing processes to support Merck's product pipeline. At that time the methyldopa manufacturing process, which even today stands as an outstanding example of what a good process should be, was just coming into its own.(*1*) The excellence and enthusiasm of the Merck Process Group was clearly evident during my interview, and I decided to make a career change before my career had begun. Thirty eight years later, I still believe that it was the right choice.

During the course of my career there have been major changes in what we do and in our responsibilities, and I am not sure that I welcome all of this, nor our response to it, with enthusiasm. In addition to our major responsibilities of designing practical syntheses and manufacturing processes, and preparing bulk drug to initiate early development, we are beset with a myriad of ever-growing regulatory requirements, endless discussions about low-level impurities, when or when not to make process changes, what are process changes, compression of program timelines, increases in drug requirements to begin development, etc., etc. Over the past ten years Process Research has emerged as a recognized field in organic chemistry, with its own journal and scientific meetings. I have always maintained that process research in the pharmaceutical industry must keep a considerable overlap with the academic chemical community. We are one of the few industrial organic chemistry enterprises that has the opportunity to do basic chemical research as a normal part of our job, and we must never let go of that opportunity. This overlap with academic research keeps us in close contact with academicians, who are keenly aware that careers in process research present wonderful opportunities for their students.

I would be remiss not to comment on our closest allies in R&D in a vehicle such as this, and I'd like to begin this section with some thoughts on medicinal chemists and medicinal chemistry routes to drug candidates. Clearly, the medicinal chemists are part of the most creative group in the pharmaceutical industry – the discovery effort, which I consider the most challenging and difficult endeavor in our industry. I stand in awe of their contributions, and I continually marvel at the complexity of structure and synthesis at which structure-activity studies are done. The era of methyl-ethyl-propyl is long since gone, and some of our carbapenem candidates reflect the heights to which discovery efforts aspire.(*2*) Such complexity in drug candidates reflects the current state of synthetic organic chemistry and organic chemists, and the armamentarium of reactions currently available. I think that process chemists often forget that the goal of medicinal chemistry is to design drug candidates, and not to design practical syntheses. Medicinal syntheses are designed for versatility, convenience and expedience. The drug candidate that results is not known when the program begins, and its initial synthesis is far from ideal by the very nature of the science. Yet process chemists persist in taking delight at setting up the medicinal synthesis of a drug candidate as a straw man to be readily pushed over with great glee. I suggest that process chemists would better serve their cause by not gloating over the inefficiencies of a first synthesis, but rather take the time to extract the best and most useful information from it and build upon it.

Second amongst our co-conspirators in R&D are the chemical engineers. We all tread with fear at the first introduction of our chemistry into the pilot plant and casting it into the hands of the chemical engineers. I had once commented to Ed Paul, the head of our chemical engineering group, that the chemists felt that the first pilot plant run for a chemical process served to confirm O'Reilly's corollary to Murphy's Law ("Whenever something can go wrong, it will go wrong."). O'Reilly noted that Murphy was an optimist. With a gleam in his eye, Ed responded that the engineers considered the first pilot plant run as the chemist's final exam! In fact, there is some truth to both views, and often problems on scale-up result from an inadequate understanding of the parameters that control our chemical processes. This is more evident than ever in today's rushed climate, where getting final drug substance as quickly as possible is often considered the major objective.

Analytical chemists and pharmaceutical scientists are also our key partners. Analytical chemistry has undergone a revolution in instrumentation, but sometimes I think this creates as many problems as it solves. The pharmaceutical scientists possibly have the most complex challenges. Despite the wonderful advances in science and technology over the past decade, much of what they do still contains a strong component of art. Regardless of the plethora of data we might acquire on a final bulk form, it is still a matter of trial and error to determine a workable drug formulation. And the only test of the response of that formulation to scale-up, is the actual scale-up.

One of my favorite topics in process research is chemical lore, and we have all fallen victim to it. Years ago we were designing and developing the first practicable synthesis of imipenem, the first carbapenem antibiotic to reach the market.(*3*) As part of this synthesis lactone **1** (Scheme 1) was solvolyzed in benzyl alcohol, which afforded a 3:1 mixture of the desired benzyl ester and the unopened lactone. This was the equilibrium ratio. The ester was cyclized to the β-lactam, which was then hydrogenolyzed to the free acid in preparation for a two-carbon chain extension. As we attempted to scale-up the benzyl alcohol solvolysis reaction, we began to recognize that such was impossible. We could not readily remove the benzyl alcohol, and as we handled the system the benzyl ester would revert to lactone. Faced with this dilemma, we wondered about solvolyzing the lactone in methanol. We were happy to note that the equilibrium ratio of the ester to lactone was now 97:3, but we were faced with the problem of hydrolyzing a methyl ester in the presence of the β-lactam. There were two parts to the established lore: everyone agreed that one could NOT successfully effect the desired hydrolysis as the β-lactam was far more reactive than the methyl ester to aqueous base; and the medicinal chemists had already demonstrated that the reaction did not work. Facing likely failure in that we could never develop the benzyl alcohol reaction into a practical process, we initially ran the hydrolysis of the methyl ester lactam with 1N NaOH in water. We were pleased that the lore was indeed wrong, and the selectivity for methyl ester hydrolysis over β-lactam opening was >100:1. Ultimately, we developed a process where the hydrolysis was done at a 30-50% concentration of the ester in water with 5N NaOH at 0°C and pH <13. The hydrolysis was exothermic and proceeded almost like a titration. After completion of the hydrolysis, the pH was adjusted to 8.5 – 9.5, and the water was evaporated at <50°C while the system was turned over to DMF in preparation for the chain extension. At 50°C and ~pH 9, the sodium salt showed no decomposition over 30 days! In fact, the $t_{1/2}$ of a 50% solution of the sodium salt at 25°C was subsequently shown to be >22 years! So my continuing advice to new process chemists is to ignore the lore of chemistry – the lore is our enemy – run the damn reaction. If it fails for the anticipated reasons, you don't have to tell the boss! If it works, the boss will think you are a genius.

Throughout my career in process research there have been a number of programs that represent, for various reasons, milestones, and I thought a recounting of some of them might be of interest to the readers of this chapter. The names of the principal contributors to these programs are listed in the references. They represent a truly excellent group of chemists, and their creativity and enthusiasm for science have been foremost in making a most enjoyable career for me. My first program at Merck was to design a new synthesis of amipramidin (**2**), a compound then being developed as a potassium-sparing diuretic. A very workable classical heterocyclic process for this drug was already on line in the pilot plant, and I was asked to design a synthesis beginning with 2-methylpyrazine (**3**) (Scheme 2). Having been trained as a synthetic organic chemist in alkaloid synthesis,(*4*) I must confess to a bit of dismay when I noted that not only did the target lack any stereo centers, but it had a most unspectacular proton NMR spectrum. Ready carbon spectra were still a dream at the time. Regardless, I began the program by looking at the chlorination of 2-methylpyrazine, on which there was essentially no literature.

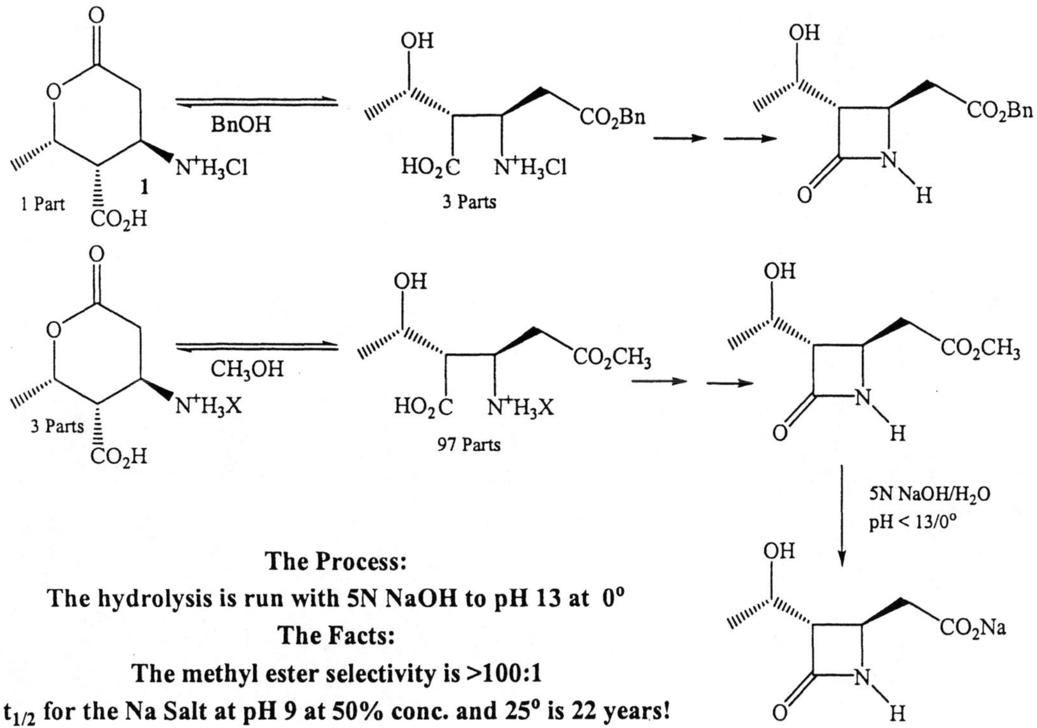

BnOH

1 Part **1** CO_2H N^+H_3Cl

HO_2C N^+H_3Cl CO_2Bn 3 Parts

OH CO_2Bn N H O

CH_3OH

3 Parts CO_2H N^+H_3X

HO_2C N^+H_3X CO_2CH_3 97 Parts

OH CO_2CH_3 N H O

5N $NaOH/H_2O$
pH < 13/0°

OH CO_2Na N H O

The Process:

The hydrolysis is run with 5N NaOH to pH 13 at 0°

The Facts:

The methyl ester selectivity is >100:1

$t_{1/2}$ for the Na Salt at pH 9 at 50% conc. and 25° is 22 years!

Scheme 1: *The lore is our enemy.*

Chlorination in water at room temperature produced two moles of nitrogen trichloride, thereby starting my career with the proverbial bang! Radical chlorination readily yielded 2-trichloromethylpyrazine (**4**), whereas ionic chlorination followed by radical chlorination afforded 2-chloro-3-dichloromethylpyrazine (**5**). Treatment of **4** with excess sodium methoxide in refluxing methanol afforded a mixture of three isomeric trimethoxypyrazines, with 2-methyl-3,5,6-trimethoxypyrazine being formed in 15% isolated yield.(*5*) I still think this is one of the most remarkable transformations that I have ever seen, and I became an instant convert to heterocyclic chemistry. I'll leave it to the reader to sort out the chemistry, but application of the principle behind the transformation to substrate **5** employing hydroxylamine in ethanol produced oxime **6**.(*6*) Clearly all of the oxidation states and substitution issues to amipramidin had been resolved, and the subsequent transformations to final product were straightforward. My new synthesis did not replace the existing route, but it did initiated my career-long love affair with heterocyclic chemistry.

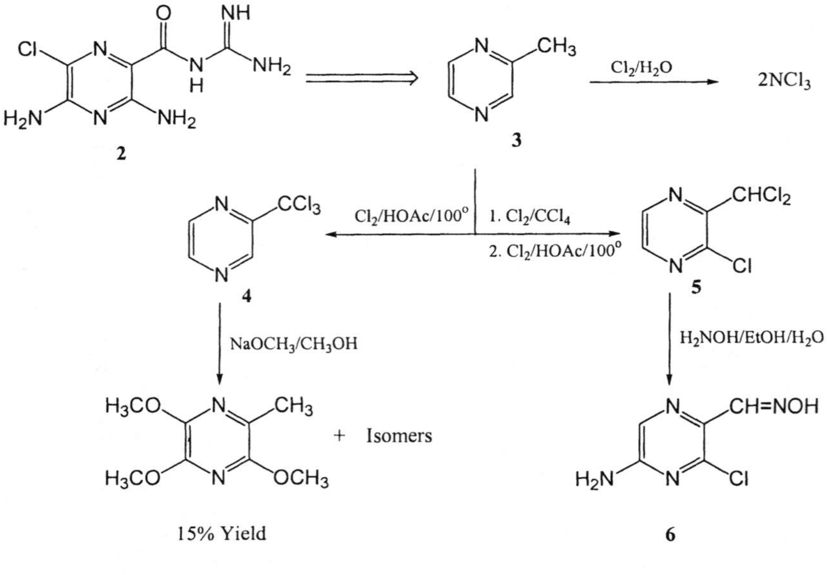

Scheme 2: *Introduction to heterocyclic chemistry.*

During the 1970's Merck, along with the rest of the industry, suffered from a weak candidate pipeline. We process chemists were asked to select or define an unusual scaffold amenable to simple and broad derivatization in an effort to define new lead candidates for the medicinal areas via broad pharmacological screening. Harkening back to my alkaloid days, I chose to prepare the tricyclic compound **8** based on reported chemistry from 4-cyano-*N*-methylpyridinium iodide (**7**) (Scheme 3).(*7*) After some failed preliminary efforts, we quickly realized that **8** was far too complex and reactive for the multitude of transformations that we wanted to employ, so we reduced the enamine double bond to form the more stable and versatile tricycle **9**. Among the numerous possible reactions that we studied, we decided to heat **9** in water. A remarkable transformation ensued, wherein our substrate, which had fourteen different carbons by ^{13}C NMR, was transformed into an isomer that showed only seven! Given that today we can do an X-ray crystal structure before lunch, this does not seem like such a difficult structural problem. However, at that time it took quite a bit of work and thought to establish **10** as the structure of the product. This remarkable rearrangement (a hydride transfer followed by ring closure of the zwitterionic intermediate) had produced a diazaditwistane with versatile functional groups just begging for further elaboration.(*8*) Recognizing the uniqueness of the structure we had produced, we joined with our associates in Chemical Engineering and prepared 10 kg of both **9** and **10** for derivatization work. I'll spare the reader the odyssey of these efforts, save to note that reaction of **10** with excess benzyl Grignard furnished bis-phenylacetyl compound **11** which was 2-3 times more potent than morphine as an analgesic.(*9*) We began a long structure-activity study that still left us with an opiate-class analgesic. The effort taught me just how difficult medicinal chemistry really is, and I went running back to process research.

The 1980's brought imipenem, the world's first carbapenem antibiotic, as the focus of our research efforts. Our goal was to define a new synthesis and ultimately process, that would supplant the one designed prior to our involvement - alluded to briefly above.(*3*) We began with azetidinone carboxylic acid (**12**), which had been previously prepared from aspartic acid in our medicinal group.(*10*) We were able to add the hydroxy ethyl group employing standard chemistry to afford **13** (Scheme 4). Oxidative decarboxylation with lead tetraacetate gave acetoxy azeditinone **14**, and we began to believe that something important was happening. Based on a paper published by Manfred Reetz wherein 1-silyloxycyclohexene was allylated with allyl acetate in the presence of zinc bromide to afford 2-allylcyclohexanone,(*11*) we quickly showed that **14** reacts with a variety of silyl enol ethers to afford the corresponding ketone products under Lewis acid catalysis.(*12*) Remembering the diazo synthon **15** developed in the penicillin approach to imipenem,(*13*) we reacted compounds **14** and **15** under Lewis acid conditions to produce **16**, a key intermediate in our earliest imipenem process. Clearly we had a new, and possibly one of the best, routes to imipenem. We published a preliminary communication on the work in 1982.(*14*) Thereafter, this route was developed to a practicable level, and combined with commercially available **14** (as the *t*-BDMS ether), this chemistry became one of the most direct approaches to imipenem. Our patent was ultimately granted in 1999 we now hold the rights to this process until 2016! (*15*)

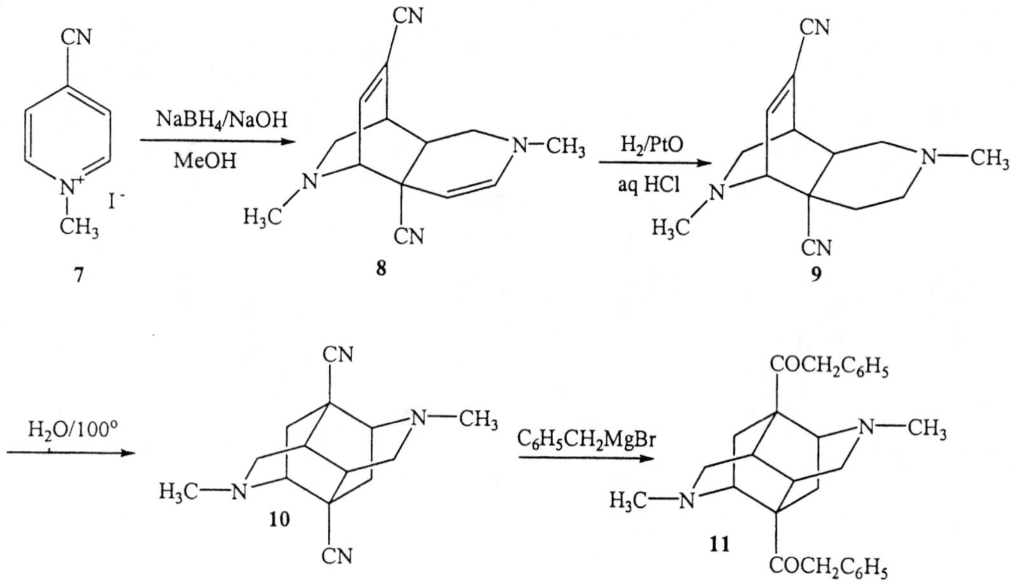

Scheme 3: *Diazaditwistane adventures.*

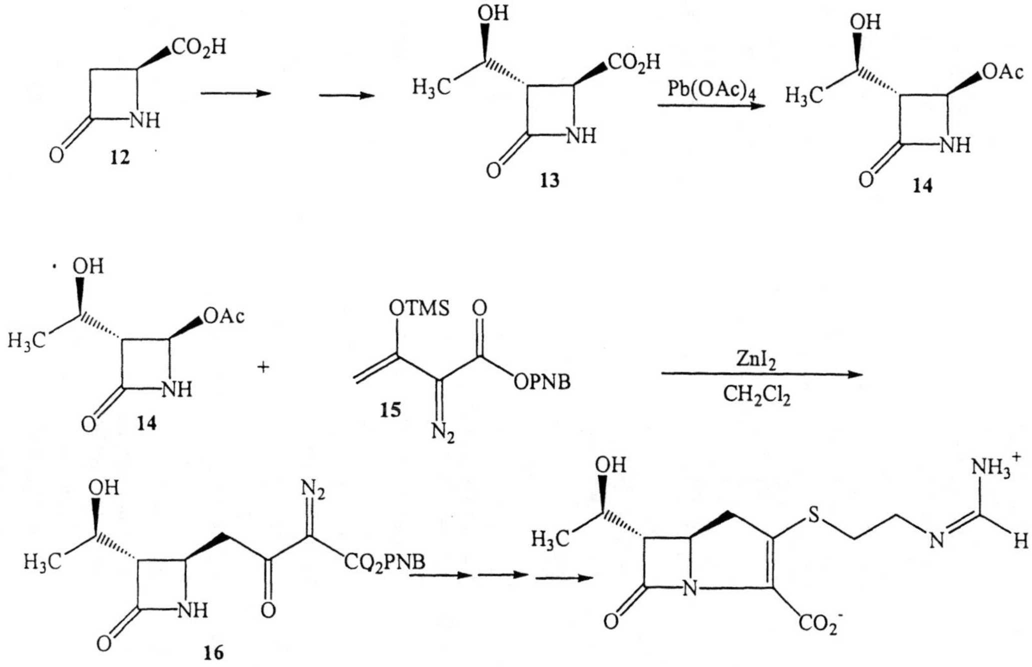

PNB = *p*-Nitrobenzyl

Scheme 4: *Acetoxy azetidinone and a practical process for imipenem.*

In the mid-1980's indacranone (**17**) came to the group, and we faced the challenge of enantiospecifically preparing a tetrasubstituted carbon center. Initially, we established the stereogenic center in intermediate **18** via an achiral phase-transfer reaction employing methyl iodide, aliquat 336, toluene and 50% sodium hydroxide. A subsequent resolution of the racemic acid provided the individual enantiomers. In thinking about how to do the same transformation enantiospecifically, we took the most simple approach initially. Rather than use the aliquat, we replaced it with simple chinchona alkaloids, thinking that these would methylate under the reaction conditions and become catalysts for chiral phase transfer chemistry. Initially our e.e.'s were less than 10%, but this provided enough encouragement that we began to make individual chinchona-derived quaternary ammonium salts. With *N*-benzylcinchonium chloride our e.e. reached ~30%. Simple electron-withdrawing *p*-substituted analogs increased the e.e., ultimately into the 60% range with the *p*-trifluoromethyl analog. Subsequent development resulted in the first enantioselective phase transfer alkylation process with the e.e. reaching 94%.(*16, 17*) This process was successfully demonstrated in the pilot plant, and supplied drug for the program until the demise of the candidate for toxicity reasons. We continued working on the chemistry and subsequently showed that the reaction mechanism is far more complex than anyone had envisioned, with the catalyst transferring into the organic phase as a 1:1 complex of the catalyst and its zwitterion. Subsequent deprotonation of the indanone by the catalyst dimer afforded the catalyst-substrate complex which involves two π-π stacking reactions and a hydrogen bond, all remarkably similar to the structure of the catalyst dimer. Methylation can only take place from the exposed face, yielding the observed isomer.(*17*)

With the recent dramatic accomplishments in phase-transfer technology developed in E.J. Corey's laboratory,(*18*) we have wondered about the correctness of this mechanism, since our catalyst does indeed *O*-methylate under our reaction conditions as an unwanted side-reaction. Could the *O*-methylated quaternary ammonium ion serve as the active catalyst, via a mechanism comparable to that proposed by Corey with his *N*-anthranceneylmethyl-*O*-allyl (or benzyl) analogs of the cinchona alkaloid catalysts as used in amino acid syntheses? In separate experiments, we had shown that the *O*-methyl catalyst does not function in our indanone case. With the reaction clearly half-order with respect to catalyst, we remain convinced as to the validity of our proposed mechanism.(*17*) The power of the phase-transfer method was revealed by a cost analysis. Although I am not at liberty to disclose the actual numbers, their relative ratios are quite interesting. If the cost of producing the racemate via the achiral phase transfer process is taken as $X/kg, the cost for producing the (*S*)-enantiomer was $3X by the resolution process. Employing the cinchona-alkaloid catalyst at <10 mol % in a chiral phase-transfer process afforded the (*S*)-isomer at a cost of $1.05X with no attempt at catalyst recycle!

Speaking of Professor Corey, our paths initially crossed during work on our carbonic anhydrase program, where we needed to use the OAB

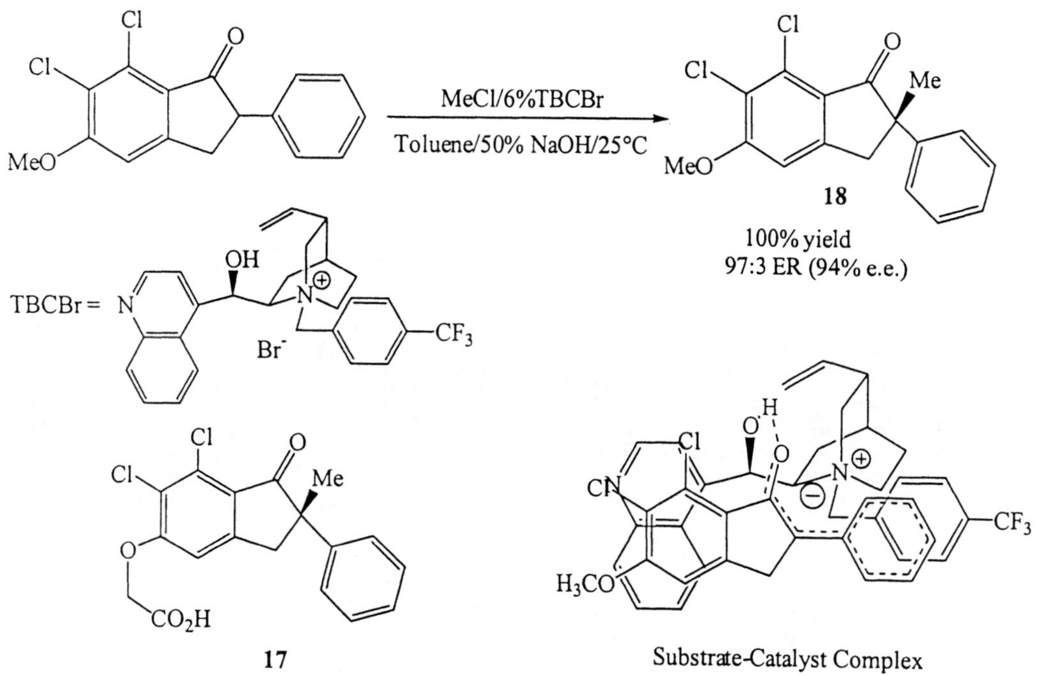

MeCl/6%TBCBr

Toluene/50% NaOH/25°C

18

100% yield
97:3 ER (94% e.e.)

TBCBr =

Br^-

17

Substrate-Catalyst Complex

Scheme 5: *Chiral catalytic phase transfer alkylation chemistry.*

(oxazaborolidine) catalyst system for enantioselective reduction of a prochiral ketone on the way to MK-417. This catalyst (**19**) had withstood the test of time as one of the most effective and elegant methods for enantioselective ketone reduction,(*19*) but clearly was applicable only to small-scale reactions based on the difficulty of purification and isolation of **19**, and its poor stability to moisture. We were interested in making this catalyst system in the pilot plant and using it on the kilogram scale. Thus, we wanted a simple method for its preparation, isolation and storage. With meticulous attention to detail, we began an investigation of these issues. This resulted in a new process for the synthesis of the OAB system, and its ready isolation as a relatively stable, crystalline BH_3 adduct. (Scheme 6),(*20*) The preparation of the isolated complex has been described in detail in Organic Synthesis.(*21*) We made more than 10kg of the catalyst in the pilot plant, and have been pleased to provide samples of it to members of the Merck Research Labs and to friends around the world.

One of the most remarkable programs that it has been my good fortune to work on was finasteride. The key problem in the medicinal chemistry route was the use of phenyl selenic anhydride to introduce the A-ring double bond. While this was a most suitable reaction at the kilogram scale, it would not be workable at the pilot scale and beyond. Ignoring the established lore, we reinvestigated the reaction of steroidal lactams with quinones and confirmed the basic research observations – no double bond introduction. This was a case where the lore was clearly correct. Noting separate papers from the Jung and Sonoda groups which reported the use of silyl enol ethers of ketones for double bond introductions in ketones that were poor substrates for quinone reactions,(*22, 23*) we started to add silylating agents to our reactions. With *N,O*-bis-(trimethylsilyl)-trifluoroacetamide (BSTFA) we achieved a 65% yield for the reaction – far better than we could do with the selenium reagent. But, things were not so copasetic: the lactam seemed to disappear faster than the product appeared. After some fairly detailed studies we established that the silyl imidate of the steroid reacted with DDQ at room temperature, presumably via an electron-transfer reaction, to afford a mixture of lactam-DDQ adducts (**20**). Subsequent thermolysis afforded the desired unsaturated product (Scheme 7). Thus, we had uncovered a completely new mechanism for quinone-mediated double bond introduction that applied to both silyl imidates and silyl enol ethers: substrate-quinone adducts form via electron-transfer reactions at room temperature followed by double bond formation under thermal conditions. These observations formed the core of a very practicable approach to finasteride, once again tying together solid new science with hardcore process research. (*24 ,25*)

During the 90's we foresaw a need for the enantiomers of **21**, one of our lead LTD_4 antagonists. The compound was proceeding in development as the racemate, despite the fact that as early as 1960 we were proceeding forward with single enantiomer drugs such as methyldopa, rather than racemates. In fact, one Merck legend states that when Max Tishler was asked whether we should proceed in development with the D,L-racemate or the single L-enantiomer of

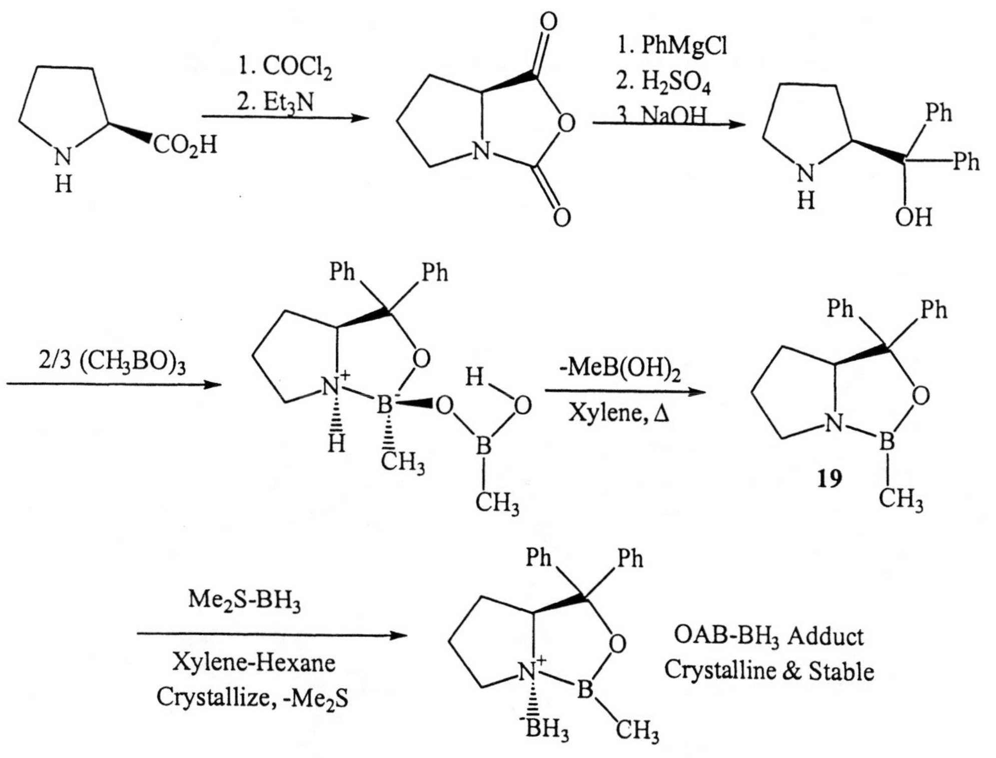

1. $COCl_2$
2. Et_3N

1. PhMgCl
2. H_2SO_4
3. NaOH

2/3 $(CH_3BO)_3$

$-MeB(OH)_2$
Xylene, Δ

19

Me_2S-BH_3
Xylene-Hexane
Crystallize, $-Me_2S$

OAB-BH_3 Adduct
Crystalline & Stable

Scheme 6: *Practical process for the preparation of Corey's OAB catalyst.*

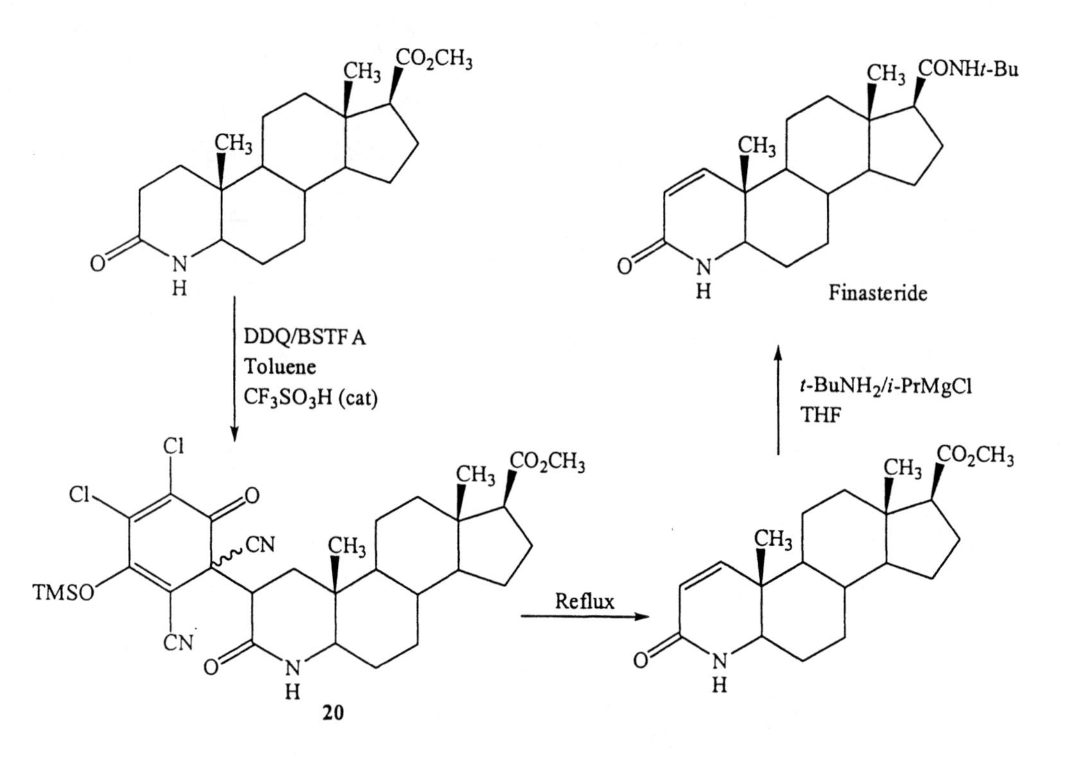

Scheme 7: *A new mechanism in quinone oxidation reactions.*

methyldopa, he responded, “Give ‘em L boys!” Back to **21** – the medicinal chemists had developed a remarkable enantioselective synthesis, but it was too complex to prepare material even at the gram level. Clearly, we needed an enzyme resolution of prochiral diester **22** (Scheme 8). Skeptics noted that such couldn’t be done as the chiral center was too far removed from the business end of the enzymatic chemistry. Bioprocess experts, on the other hand, encouraged us to pursue this avenue of research despite the lack of literature precedent. We took on the challenge and the results were everything we could expect from this gambit in bioprocess. Using a lipase from Pseudomonas a >98% e.e. was achieved in the resolution at 90% chemical yield to afford (*S*)-ester acid **23**.(*26*) From this compound we could make either isomer of the target by direct amination or acid activation, amination and hydrolysis. Thus, we were able to provide quantities of both enantiomers for serious biological evaluation for the first time, and the practicality of the work was subsequently demonstrated in numerous pilot plant campaigns. Ali Shafiee has noted to me many times that this example is still cited at current symposia on enzymatic reactions as a prime example of what can be done with biotransformations.(*27*)

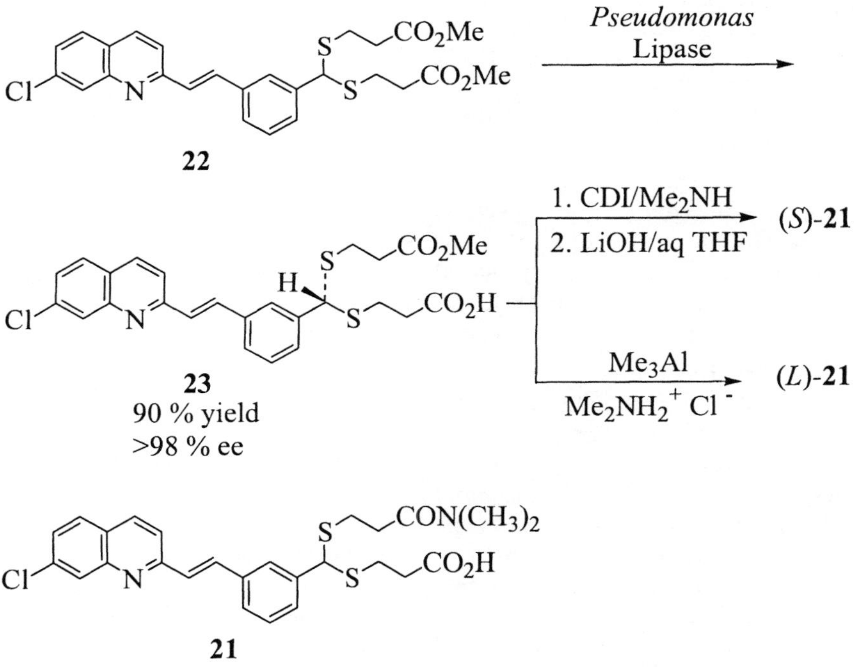

CO_2Me CO_2Me Cl N S S *Pseudomonas* Lipase

22

1. CDI/Me_2NH 2. LiOH/aq THF → (*S*)-**21**

CO_2Me CO_2H H Cl N S S

23
90 % yield
>98 % ee

Me_3Al / $Me_2NH_2^+ Cl^-$ → (*L*)-**21**

$CON(CH_3)_2$ CO_2H Cl N S S

21

Scheme 8: *A gambit in biocatalysis.*

Possibly one of the most exciting programs that I have ever worked on is the Merck RT (reverse transcriptase) program which ultimately led to efavirenz. The program began with target **24**, which was originally made as the reacemate and then separated into its enantiomers by chiral derivatization and chromatography.(*28*) Initially we developed a remarkable approach to this molecule by the quinine lithium alkoxide-mediated addition of the lithium acetylide to the imine precursor protected on nitrogen by the anthranylmethyl group. This reaction at –25°C afforded a solution e.e. of 97% in 84% chemical yield at the 1.8 kg scale (Scheme 9).(29)

As often happens, the medicinal chemists then came up with a more potent inhibitor (**25**, efavirenz),(*28*) and asked us to focus instead on it. Keeping the earlier concept in mind, we were able to achieve an e.e. in the 80's by preforming a mixture of lithium cyclopropyl acetylide and the alkoxide of ephedrine derivative **26** in THF and then adding ketone **27** which had a *p*-methoxybenzyl protecting group (PMB) on nitrogen. Remembering that we had observed a remarkable temperature response in the imine addition reaction (the e.e. decreased at higher AND lower temperatures) and that the nature of lithium aggregates changes with temperature, we had the audacity to warm the THF solution of the lithium acetylide and lithium ephedrine alkoxide to 0°C, prior to cooling to <-50°C for the ketone addition. Under these conditions, we achieved a 98+% e.e. at 98% conversion (Scheme 10).(*30*) We later asked Dave Collum of Cornell to join us in the study of this reaction, and Dave showed by ^{6}Li NMR studies that the equilibration produces a cubic tetramer composed of two acetylides and two alkoxides (**28**).(*31*) Dave, of course, got all of us thinking about lithium aggregation and ^{6}Li-NMR studies, and in some NMR work of our own we showed that subsequent reaction of this tetramer with the PMB-protected ketone produces a new cubic tetramer composed of one acetylide, two ephedrine alkoxides and one product alkoxide with the product at 98+% e.e. (*32*) As to our interactions with Dave Collum, I can only say that he is a prince among chemists, and he has changed the ways in which we view all lithium anion reactions.

Wondering if the job of a process chemist is ever done, we decided to look at development of a process that would not require PMB protection and deprotection, and loss of one mole of acetylide to the product tetramer. We ended up moving to the zinc manifold recognizing problems of the unprotected substrate with strong bases and the product with strong Lewis acids. The new process that resulted was even more remarkable than the lithium acetylide process (Scheme 11). A zinc reagent mixture was prepared from diethyl zinc and the Grignard of the acetylide in the presence of the SAME ephedrine mediator and a simple achiral alcohol, and addition of the unprotected ketone at 0-25°C produced the desired product in 98% yield and 99.3% e.e. (*33*) Attempted mechanistic studies were not productive; however, the results of the reactions were consistent regardless of how we made up the reagent mixture. Clearly, we had a very dynamic mixture of zinc species, which only afforded the desired product.

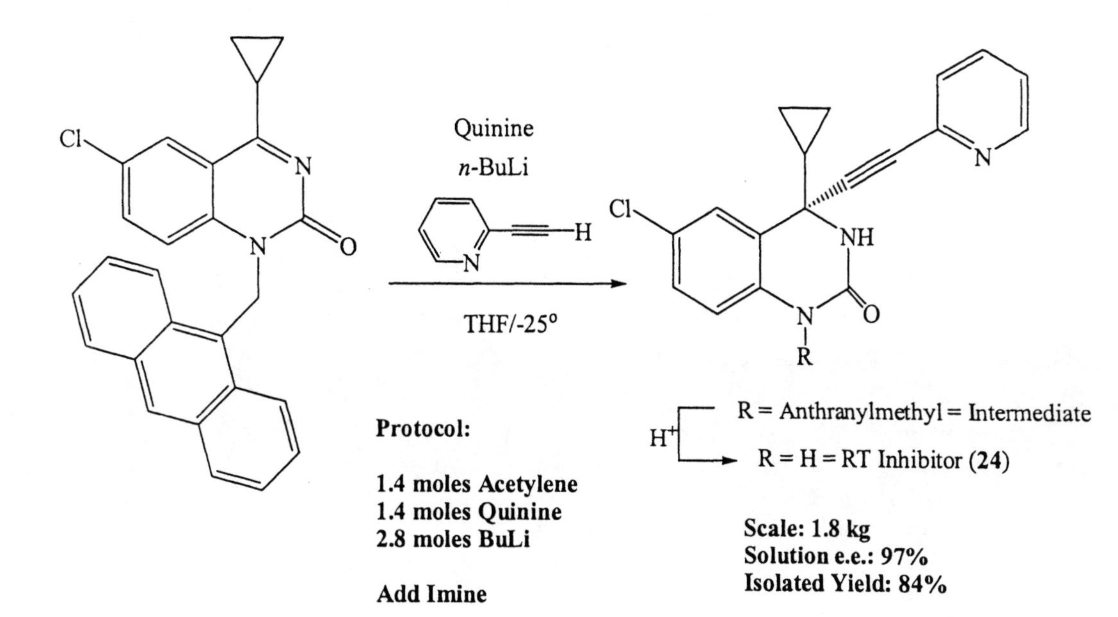

Figure 9*. Our first enantioselective acetylide addition.*

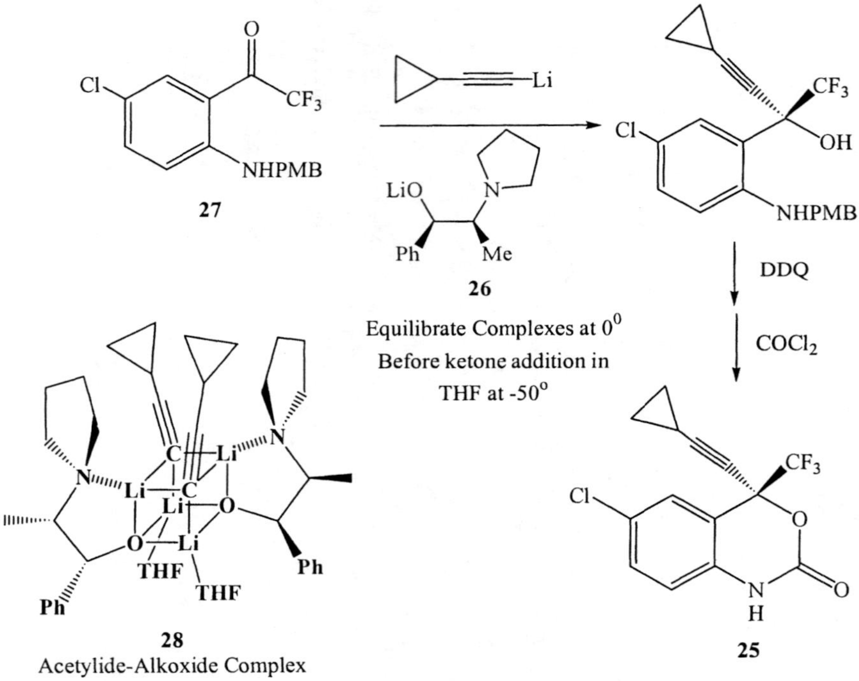

27

26

Equilibrate Complexes at 0^0
Before ketone addition in
THF at -50°

DDQ

$COCl_2$

28
Acetylide-Alkoxide Complex

25

Scheme 10: *Efaverinz and understanding enantioselective lithium alkoxide-mediated lithium acetylide additions.*

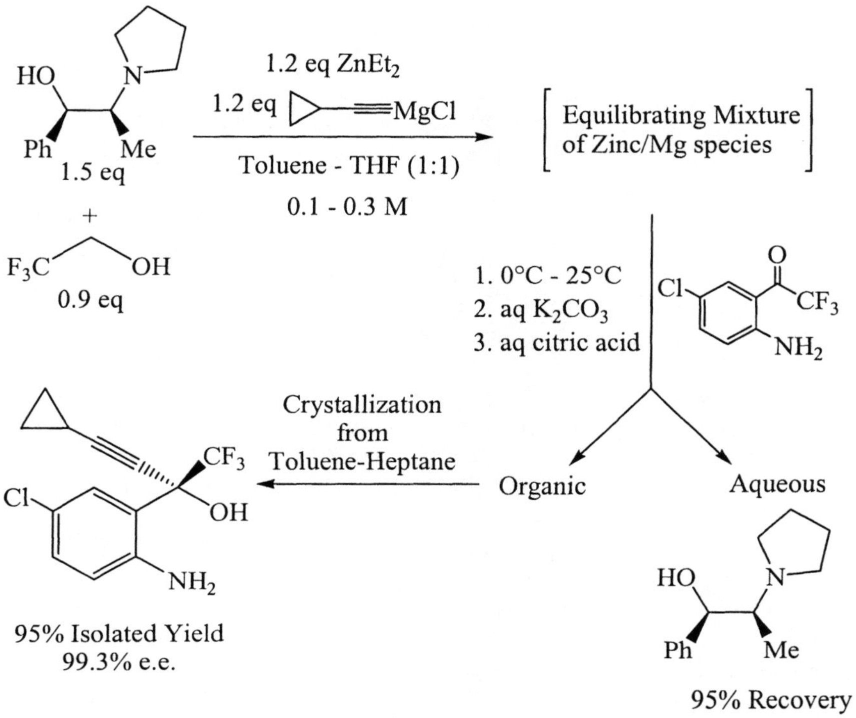

***Scheme 11**: Enantioselective zinc acetylide addition applied to efavirenz.*

There are many more programs that I could highlight in this presentation were there more space. Other work of which it has been my good fortune to be a small part includes work on the mechanism of the Mitsunobu reaction, enantioselective ketene additions, numerous aspects of avermectin chemistry, novel oxidation methods, immunosuppressant synthesis and novel carbene rearrangements. Again, I owe a tremendous debt to all of my coworkers on these programs – it is they who have kept me sailing in this marvelous journey in chemistry and chemical development. I think that all of this work clearly shows how process research and fundamental chemical research are closely related. Additional stories are being readied for telling, and I am looking forward to the pending publications from our younger staff members. One from Jerry Murry appears in this very volume. To those young chemists who decide to work in process research, I can only wish you 'calm seas and a prosperous voyage' in what I am sure will be an exciting career.

References:

1. Firestone, R. A.; Reinhold, D. F.; Gaines, W. A.; Chemerda, J. C.; Sletzinger, M. *J. Org. Chem.* **1968**, *33*, 1213-1218. Midler, Jr, M. U. S. Patent 3,892,539, 1975.
2. Humphrey, G. R.; Miller, R. A.; Pye, P. J.; Rossen, K.; Reamer, R. A.; Maliakal, A.; Ceglia, S. S.; Grabowski, E. J. J.; Volante, R. P.; Reider, R. J. *J. Am. Chem. Soc.* **1999**, *121*, 11261-11266, and references cited therein.
3. Melillo, D. G.; Cvetovich, R. J.; Ryan, K. M.; Sletzinger, M. *J. Org. Chem.* **1986**, *51*, 1498-1504 and references cited therein.
4. Grabowski, E. J. J.; Autrey, R. L. *Tetrahedron* **1969**, *25*, 4315-4330, and references cited therein.
5. Grabowski, E. J. J.; Tristram, E. W.; Tull, R. J.; Pollak, P. I. *Tetrahedron Lett.* **1968**, *9*, 5931-5934.
6. Grabowski, E. J. J.; Tristram, E. W.; Tristram, E. W. U. S. Patent 3,625,944, 1971.
7. Liberatore, F.; Casini, A.; Carelli, V. *Tetrahedron Lett.* **1971**, *12*, 2381-2385.
8. Ten Broeke, J.; Douglas, A. W.; Grabowski, E. J. J. *J. Org. Chem.* **1976**, *41*, 3159-3163.
9. Fisher, M. H.; Grabowski, E. J. J.; Patchett, A. A.; ten Broeke, J.; Flataker, L. M.; Lottie, F. M.; Robinson, F. M. *J. Med. Chem.* **1977**, *20*, 63-66.
10. Salzmann, T. N.; Ratcliff, R.; Christensen, B. G.; Bouffard, F. A. *J. Am. Chem. Soc.* **1980**, *102*, 6161-6163.
11. Reetz, M. T.; Hüttenhain, S.; Hübner, F. *Synth. Comm.* **1981**, *11*, 217-222.
12. Reider, P. J.; Rayford, R.; Grabowski, E. J. J. *Tetrahedron Lett.* **1982**, *23*, 379-382.
13. Karady, S.; Amato, J. S.; Reamer, R. A.; Weinstock, L. M. *J. Am. Chem. Soc.* **1981**, *103*, 6765 –6767.
14. Reider, P. J.; Grabowski, E. J. J. *Tetrahedron Lett.* **1982**, *23*, 2293-2296.
15. Reider, P. J.; Grabowski, E. J. J. U. S. Patent 5,998,612, 1999.
16. Dolling, U.-H.; Davis, P,; Grabowski, E. J. J. *J. Am. Chem. Soc.* **1984**, *106*, 446-447.
17. Hughes, D. L.; Dolling, U.-H.; Ryan, K. M.; Schoenewaldt, E. F.; Grabowski, E. J. J. *J. Org. Chem.* **1987**, *52*, 4745-4752.
18. Corey, E. J.; Xu, F.; Noe, M. C. *J. Am. Chem. Soc.* **1997**, *119*, 12414-12415.
19. Corey, E. J.; Bakshi, R. K.; Shibata, S. *J. Am. Chem. Soc.* **1987**, *109*, 5551-5553.
20. Mathre, D. J.; Thompson, A. S.; Douglas, A. W.; Hoogsteen, K.; Carroll, J. D.; Corley, E. G.; Grabowski, E. J. J. *J. Org. Chem.* **1993**, *58*, 2880-2888.

21. Xavier, L. C.; Mohan, J. J.; Mathre, D. J.; Thompson, A. S.; Carroll, J. D.; Corley, E. G.; Desmond, R. *Organic Synthesis*, **1998**, *Collective Volume IX*, 676-688.
22. Jung, M. E.; Pan, Y.-G.; Rathke, M. W.; Sullivan, D. F.; Woodbury, R. P. *J. Org. Chem.* **1977**, *42*, 3961-3963.
23. Ryv, I.; Murai, S.; Hatayame, Y.; Sonoda, N. *Tetrahedron Lett.* **1978**, *19*, 3455-3458.
24. Bhattacharya, A.; DiMichele, L. M.; Dolling, U.-H.; Douglas, A. W.; Grabowski, E. J. J. *J. Am. Chem. Soc.* **1988**, *110*, 3318-3319.
25. Bhattacharya, A.; DiMichele, L. M.; Dolling, U.-H.; Grabowski, E. J. J.; Grenda, V. J. *J. Org. Chem.* **1989**, *54*, 6118-6120.
26. Hughes, D. L.; Bergan, J. J.; Amato, J. S.; Reider, P. J.; Grabowski, E. J. J. *J. Org. Chem.* **1989**, *54*, 1787-1788.
27. Mr. Ali Schafiee, personal communication.
28. Young, S. D.; Britcher, S. F.; Tran, L. O.; Payne, L. S.; Lumma, W. C.; Lyle, T. A.; Huff, J. R.; Anderson, P. S.; Olsen, D. B.; Carroll. S. S.; Pettibone, D. J.; O'Brien, J. A.; Ball, R. G.; Balani, S. K.; Lin, J. H.; Chen, I.-W.; Schleif, W. A.; Sardana, V. V.; Long, W. J.; Byrnes, V. W.; Emini, E. A. *Antimicrob. Agents Chemother.* **1995**, *39*, 2602.
29. Huffman, M. A.; Yasuda, N.; DeCamp, A. E.; Grabowski, E. J. J. *J. Org. Chem.* **1995**, *60*, 1590-1594.
30. Thompson, A. S.; Corley, E. C.; Huntington, M. F.; Grabowski, E. J. J. *Tetrahedron Lett.* **1995**, *36*, 8937-8940.
31. Thompson, A.; Corley, E. G.; Huntington, M. F.; Grabowski, E. J. J.; Remenar, J. F.; Collum, D. B. *J. Am. Chem. Soc.* **1998**, *120*, 2028-2038.
32. Xu, F.; Reamer, R. A.; Tillyer, R.; Cummins, J. M.; Grabowski, E. J. J.; Reider, P. J.; Collum, D. B.; Huffman, J. C. *J. Am. Chem. Soc.* **2000**, *122*, 11212-11218.
33. Tan, L.; Chen, C.; Tillyer, R. D.; Grabowski, E.J.J.; Reider, P. J. *Angew. Chem. Int. Ed.* **1999**, *38*, 711-713.

Chapter 2

Synthesis of Cell Adhesion Inhibitors via Crystallization-Driven Dynamic Transformations

Magnus Eriksson[1], Vittorio Farina[1], Suresh Kapadia[1], Elio Napolitano[1,2], and Nathan K. Yee[1]

[1]**Department of Chemical Development, Boehringer Ingelheim Pharmaceuticals, Ridgefield, CT 06877**
[2]**Dipartimento di Chimica Bioorganica e Biofarmacia, via Bonanno 33, 56100 Pisa, Italy**

Two approaches to BIRT-377 (**1**) are discussed. The focus is the stereoselective synthesis of *trans*-imidazolidinones such as **16** and *cis*-oxazolidinones such as **29d**. The kinetic and thermodynamic factors governing the *cis/trans* selectivity in the formation of **16** and **29d** were studied, and it was found that neither can assure complete selectivity in favor of either form. It was then found in both cases that a crystallization-driven dynamic transformation can produce, in a very efficient manner, the desired *cis* isomer in virtually 100% selectivity in the case of **29d**, whereas only the *trans* isomer is obtained in the case of **16**. Self-regeneration of stereocenters is then applied to the alkylation of the enolates derived from **16** and **29d** with *p*-bromobenzyl bromide, followed by routine transformations, to produce **1** in >99.9% ee *via* two separate processes.

Introduction

BIRT 377 (**1**) is the first prototype of an exciting class of small molecules, which are capable of inhibiting cell adhesion by blocking protein-protein interaction between LFA-1 and ICAM-1.[*1, 2*]

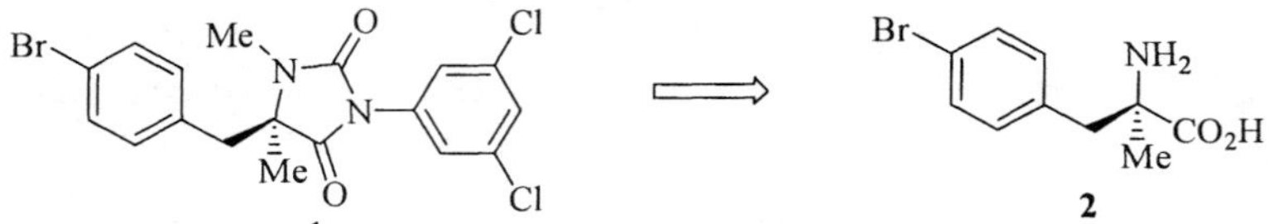

The structure of the target compound is simple and the only synthetic challenge is presented by the presence of a chiral quaternary center. Retrosynthetically, **1** can be derived from a chiral α-methyl amino acid (**2**), which belongs to a class of compounds which is readily available by a number of approaches.[*3*] The discovery route (Scheme 1) did not seek to address enantioselectivity and relied instead on chromatographic resolution to separate BIRT 377 from its distomer.

When **1** was selected for scale-up, we were required to prepare very rapidly a moderate amount of material for initial studies (*ca.* 500g) by what is commonly referred to as an "expedient" synthesis.[*4*] Given the limited time frame and the pressure on the process chemist to deliver quickly an initial batch of the drug candidate, this route is usually the discovery route after suitable adjustments, that is, in our specific case, the introduction of a resolution step. *At the same time*, we sought to develop a "scalable" route, *i.e.* one that could eventually deliver quickly and cheaply large amounts of material that will be required later on.

Scheme 1

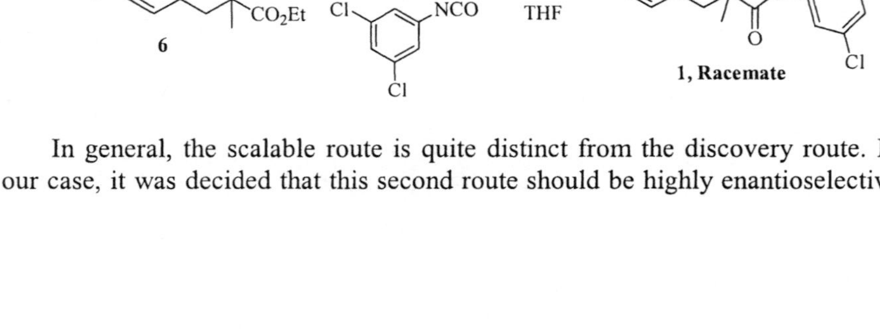

NH_3Cl; CO_2Et; **3**; 1. TEA, CH_2Cl_2; 2. *t*-BuCHO, $MgSO_4$, CH_2Cl_2; N; CO_2Et; **4**; Br, Br; *t*-BuOK, THF; Br; N; CO_2Et; **5**; 1N HCl; Br; NH_2; CO_2Et; **6**; DMSO, heat; Cl; NCO; Cl; MeI, LDA; THF; Br; N; O; N; Cl; O; Cl; **1, Racemate**

In general, the scalable route is quite distinct from the discovery route. In our case, it was decided that this second route should be highly enantioselective and, given the current status of the field of enantioselective synthesis, it should not have to rely on resolutions, which would waste half of a key intermediate.

The Expedient Route

Given literature precedents, [*5, 6*] it seemed logical that intermediate **6** should be amenable to enzymatic resolution. Indeed, Lipase L (from Amano) provided a good yield of resolved **7** in high enantiomeric excess (>99%) after a few days at rt, although it proved sluggish even at the optimum pH (6.4). Following closely the discovery route, **1** was then prepared in reasonable purity, with only about 0.2% impurities plus *ca.* 0.7% of the undesired enantiomer. This quality was more than satisfactory for the initial toxicology study (**Scheme 2**). Obviously, the synthesis illustrated in Schemes 1 and 2 leaves much to be desired from the standpoint of practicality, although short (7 steps) and fairly high yielding (26.4% from *L*-alanine). Among the few problems, the enzymatic step had very low throughput and employed large quantities of enzyme. Although this could be recycled, the 50% maximum yield achievable by this process makes the approach utterly inefficient.

The discovery synthesis starts with enantiomerically pure (*L*)-alanine. Unfortunately, the enantiomeric purity is lost during the synthetic manipulations. We reasoned that, if the stereochemistry at the tertiary carbon could be somehow efficiently preserved, then cheap alanine derivatives could be very practical starting materials. This approach led to two distinct highly efficient processes, which will form the subject of the next sections.

Asymmetric Synthesis via Imidazolidinones[*7*]

The key structural feature of BIRT-377 (**1**) is the *N*-aryl-substituted-hydantoin bearing a quaternary stereogenic center. In our retrosynthetic analysis, the hydantoin ring can be synthesized by cyclization of the corresponding acyclic α-substituted amino acid amide (**9**), which could be derived from (*D*) or (*L*)-alanine.

In the literature, a variety of methodologies have been reported for the asymmetric synthesis of quaternary α-amino acids.[*3*] Among these methods, Seebach's self-regeneration of stereocenters [*8*] for the preparation of α-substituted amino acid derivatives (Scheme 3) was most attractive to us. The principal reasons are the ready availability of the required chiral imidazolidinones (up to 90% ds), the stability of the corresponding enolates at higher temperature (up to 0 °C), and the predictability of the stereochemical outcome.

Scheme 2

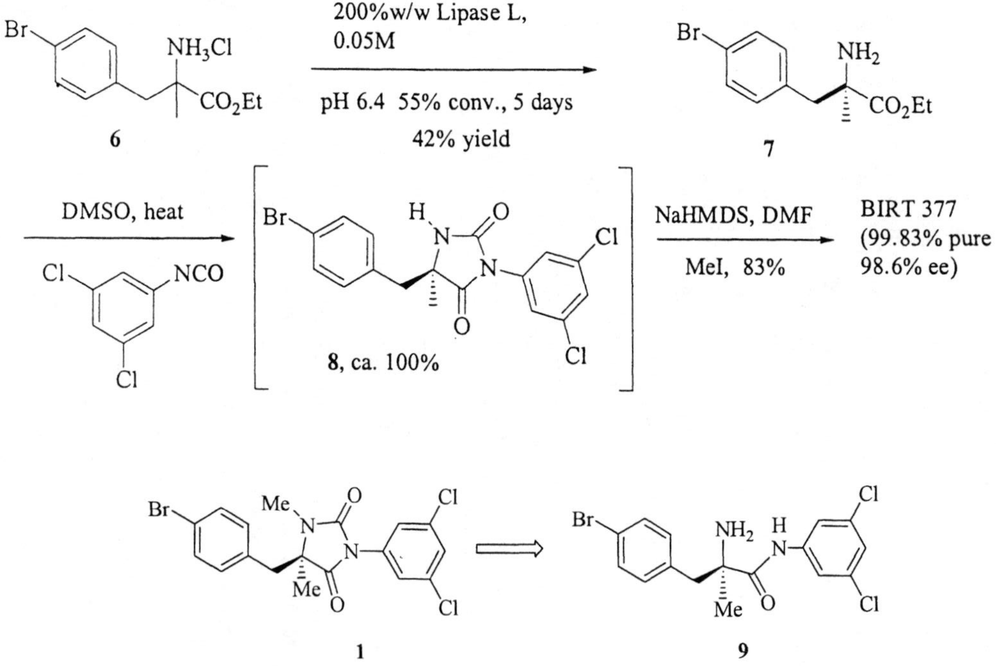

Br
NH_3Cl
CO_2Et
6

200%w/w Lipase L,
0.05M
pH 6.4 55% conv., 5 days
42% yield

Br
NH_2
CO_2Et
7

DMSO, heat
Cl
NCO
Cl

Br
H
N
O
N
Cl
O
Cl
8, ca. 100%

NaHMDS, DMF
MeI, 83%

BIRT 377
(99.83% pure
98.6% ee)

Br
Me
N
O
N
Cl
Me
O
Cl
1

Br
NH_2
H
N
Cl
Me
O
Cl
9

Scheme 3

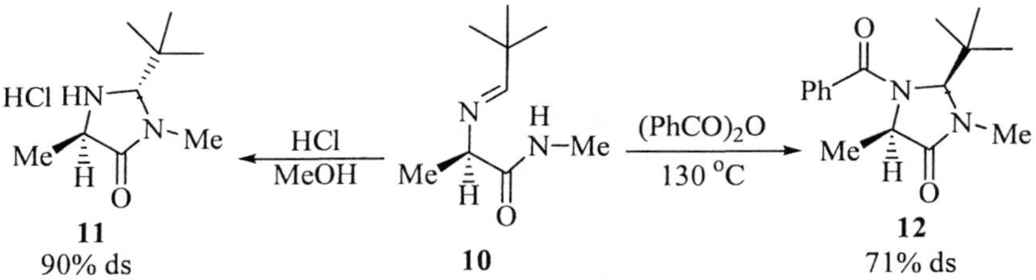

HCl MeOH ← **10** → $(PhCO)_2O$, 130 °C

11 90% ds | **10** | **12** 71% ds

Surprisingly, the method has not been widely used, especially for industrial large-scale production. This is probably due to the harsh conditions (aq. HCl, 150-220 °C) required to hydrolyze the resulting 5,5-disubstituted imidazolidinone, [*9-19*] as well as the modest diastereoselectivities [Scheme 3, **11** (90% ds), **12** (71% ds)] observed in the formation of the imidazolidinones.[*9*] During our survey of the literature, it was also found that the diastereoselective formation of either the *trans* or *cis* imidazolidinones by Seebach's method was reported only for the α-amino *N*-methyl amides as substrates, whereas other substitutions (such as *N*-aryl) on the amide nitrogen atom have not been studied.[*3, 8*] Therefore, the utilization of Seebach's principle to synthesize α-substituted amino *N*-aryl amide **9** is intriguing in relation to the possible stereochemical outcome.

Herein, we describe an efficient enantiospecific synthesis of BIRT-377 (**1**) based on a modification of Seebach's strategy in which we extend the original protocol to achieve complete (>99.9%) overall diastereoselectivity.

The synthesis of BIRT-377 (**1**) is outlined in Scheme 4. The commercially available (D)-*N*-Boc-alanine (**13**) reacted with 3,5-dichloroaniline via a mixed anhydride intermediate (*i*-BuOCOCl, *N*-methylmorpholine, -10 °C to rt, THF) to give amide **14**. Deprotection of the crude amide **14** with trifluoroacetic acid in dichloromethane afforded amino *N*-aryl amide **15** in 92% yield over two steps. This crude product was pure enough to carry on for next step without any purification.

In the early laboratory studies, amino amide **15** was treated with pivalaldehyde in refluxing pentane as described in Seebach's original procedure. A crystalline solid was directly formed from the reaction mixture and identified as the desired *trans* imidazolidinone **16** as a single diastereomer in 74% yield. This observation is in contrast with Seebach's case for the corresponding amino acid *N*-methyl amide. Seebach reported that the acyclic Schiff base intermediate was actually obtained in this step and then cyclized only when treated with either HCl in MeOH at 0 °C or $(PhCO)_2O$ at 130 °C (Scheme 3). Later (Scheme 5), we found that a mixture of *trans*/*cis* imidazolidinones **16**/**20** and Schiff base **19**, produced by treating amino amide **15** with pivalaldehyde in toluene or dichloromethane, was completely converted to the pure *trans* isomer **16** upon

Scheme 4

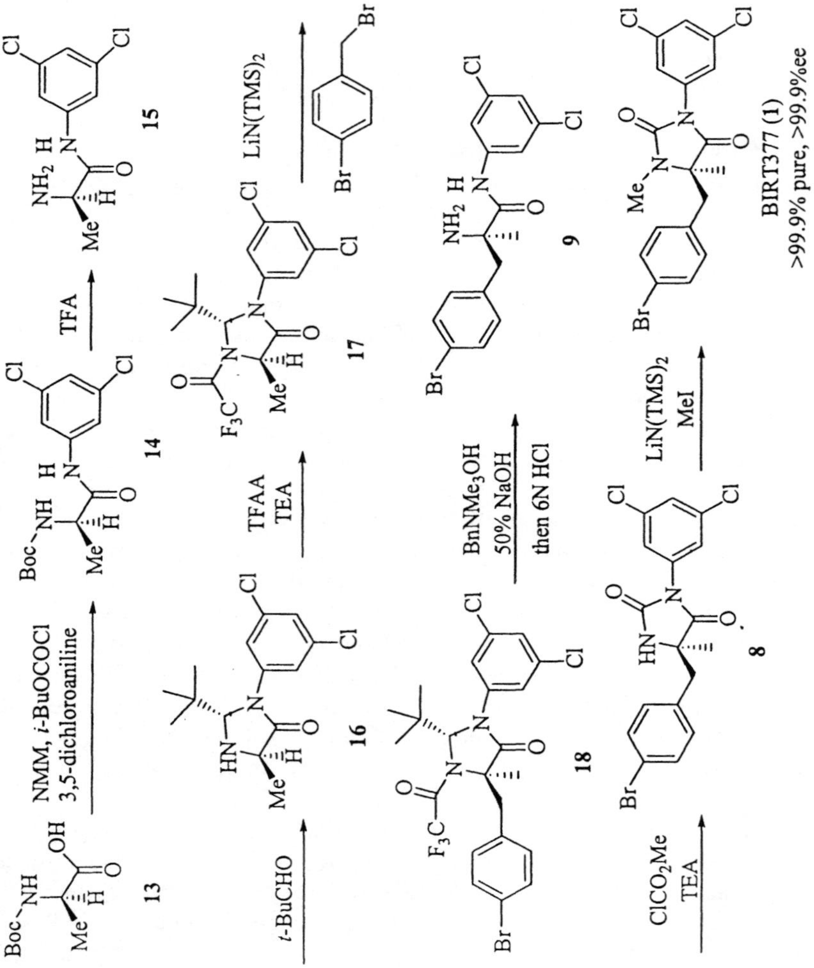

13 → 14: NMM, *i*-BuOCOCl, 3,5-dichloroaniline

14 → 15: TFA

15 → 16: *t*-BuCHO

16 → 17: TFAA, TEA

17 → 18: $LiN(TMS)_2$, 4-bromobenzyl bromide

18 → 9: $BnNMe_3OH$, 50% NaOH, then 6N HCl

9 → 8: $ClCO_2Me$, TEA

8 → 1: $LiN(TMS)_2$, MeI

BIRT377 (**1**)
>99.9% pure, >99.9%ee

crystallization. This occurred either when the neat mixture stood over a period of time, or when the mixture of **16/20/19** was crystallized from a non-polar solvent like pentane. These experiments suggest that the crystallinity of *trans* **16** is the driving force for the stereospecific formation of a single diastereomer.

After protection of **16** with $(CF_3CO)_2O$ (Et_3N, 0 °C to rt, CH_2Cl_2, 98% yield), crude **17** in THF was deprotonated with $LiN(TMS)_2$ at –30 to -20 °C and then the resulting enolate was alkylated at -30 °C to 0 °C with 4-bromobenzyl bromide from the opposite face of the *t*-butyl group, to give **18** as a single diastereomer in 96% yield.

Hydrolysis of dialkylated imidazolidinone **18** was somewhat problematic, as expected. The substrate remained intact under most of the traditional hydrolysis conditions (aq. HCl/MeOH, reflux; aq. NaOH/MeOH, reflux; or H_2NNH_2/MeOH, reflux). After considerable effort, a practical one-pot two-step hydrolysis procedure was developed. The trifluoroacetamide group of **18** was first hydrolyzed (1.5 eq. $BnMe_3NOH$, 2.0 eq. 50% NaOH, rt to 40 °C, dioxane) to give a mixture of the corresponding partially hydrolyzed *N*-unsubstituted acetal of **18**, Schiff base of **9**, and **9** itself. Subsequent direct addition of 6N HCl to the above mixture resulted in complete hydrolysis to afford amino amide **9** in quantitative yield.

Treatment of crude **9** with methyl chloroformate in the presence of triethylamine gave crude hydantoin **8** in 91% yield. Methylation [$LiN(TMS)_2$, MeI, DMF, rt] of **8** followed by a single recrystallization of the crude product from EtOAc/hexane afforded BIRT-377 (**1**) in 74% yield. The drug substance prepared by this method possessed excellent chemical and enantiomeric purity (both >99.9% by HPLC).

In summary, an efficient enantiospecific synthesis of the LFA-1 antagonist BIRT-377 (**1**) has been achieved in 43% overall yield in 8 steps. The key transformations involve the stereospecific formation of the crystalline *trans* imidazolidinone **16**, subsequent alkylation, and the efficient hydrolysis of **18**. It should be noted that the crude intermediates were used directly in all steps in this synthetic scheme and there was no purification step needed in the entire sequence, except for the crystallization of **16**. A single crystallization of the final product gave the drug substance in high purity. This process is practical, robust, and cost-effective; it has been successfully implemented in the pilot plant to produce multi-kilogram quantities of BIRT-377 (**1**).

In order to reduce the overall cost of the process, replacement of pivalaldehyde by the more available and less expensive isobutyraldehyde was desired. Even though such change may seem trivial, *it has not been previously documented in literature that aldehydes bearing α-hydrogens can be used in this protocol.* Fortunately, with a minor modification of the current procedure, the incorporation of isobutyraldehyde in the process was successful and produced similar results as those described above (Scheme 6).[*20*]

Scheme 5

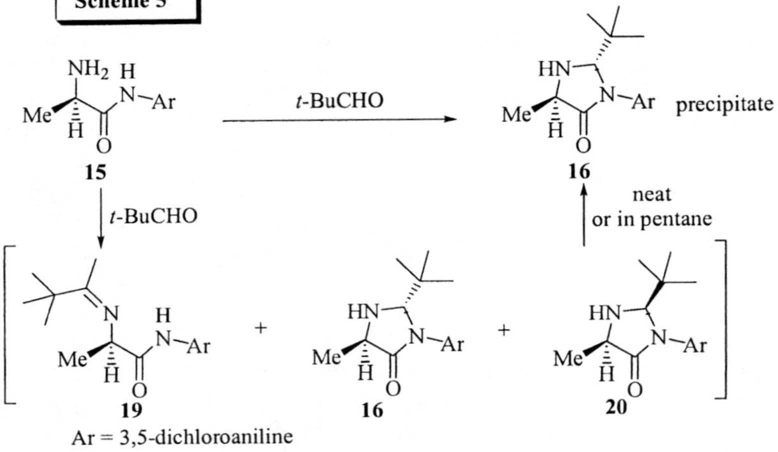

NH_2 H
N–Ar
Me H O
15

t-BuCHO

HN
N–Ar precipitate
Me H O
16

t-BuCHO

neat
or in pentane

N H
N–Ar
Me H O
19

+

HN
N–Ar
Me H O
16

+

HN
N–Ar
Me H O
20

Ar = 3,5-dichloroaniline

Scheme 6

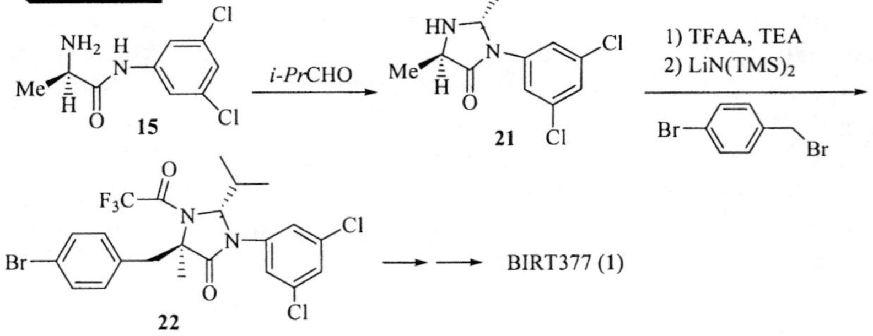

NH_2 H
N
Cl
Cl
Me H O
15

i-PrCHO

HN
N
Cl
Me H O
21 Cl

1) TFAA, TEA
2) $LiN(TMS)_2$
Br
Br

O
F_3C
N
N
Cl
Br
O
Cl
22

BIRT377 (**1**)

Asymmetric Synthesis via Oxazolidinones

The synthesis of the amino amide **9** via alkylation/deprotection of an imidazolidinone is efficient and provides BIRT 377 in high purity and in relatively few steps.[*7*] However, the sequence involves a hydrolysis step which generally requires forcing conditions, and involves 3 steps that only achieve protection and deprotection of functional groups. We reasoned that if we could use an oxazolidinone template for a similar alkylation reaction, we should be able to use a milder deprotection procedure to isolate a derivative of the quaternary amino acid such as **23**. On the basis of the literature, this would allow us to use the cheaper (L)-alanine. In addition, protection/deprotection strategies would be minimized.

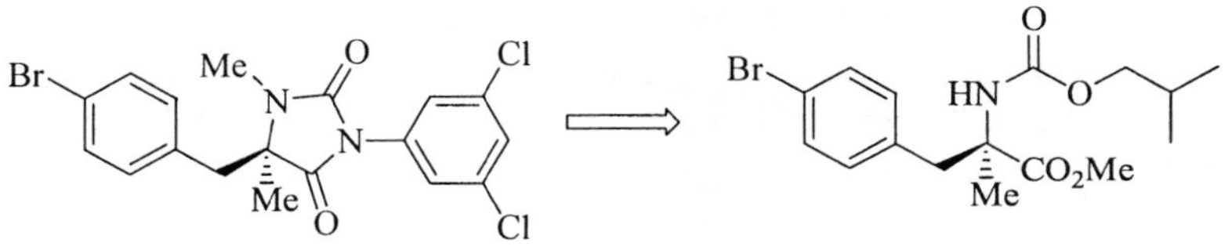

1 ⟹ **23**

There are several methods to prepare oxazolidinones from α-amino acids, for the use as templates in alkylation reactions.[*3*] Our early investigations focused on the protocol by Seebach based on acylation of imine salts of amino acids, [*8*] but we encountered problems associated with the heterogeneous formation of imine salts on larger scale. A method was then developed by which commercially available (*L*)-*N*-Cbz-alanine (**24**) is reacted with benzaldehyde dimethylacetal(**25**) in the presence of $SOCl_2$ and $ZnCl_2$ to provide the oxazolidinone **26** with a predominant *cis*-relationship (*cis:trans ca.* 5-15:1) between the phenyl group and the α-methyl group, as shown in Scheme 7.[*21*]

Scheme 7

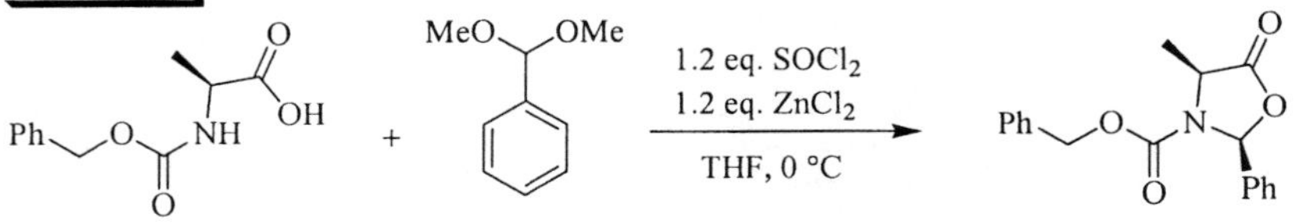

24 + **25** → (1.2 eq. $SOCl_2$, 1.2 eq. $ZnCl_2$, THF, 0 °C) → **26**; 65% yield

Although this method provides reasonable yield of **26**, a recrystallization is needed to reach sufficient diastereomeric purity before the alkylation reaction.

During this work, we came across a report where catalytic amounts of $SOCl_2$ had been used to promote oxazolidinone formation from glycine.[*22*] Because we used stoichiometric amounts of $SOCl_2$, this suggested that perhaps an acid chloride could be an intermediate in this transformation. However, acid chlorides from *N*-acyl amino acids are difficult to prepare and are known to be unstable. Buckley *et al.* have shown that acid chlorides from *N*-acyl amino acids undergo racemization via azlactone formation.[*23*] This reaction is much slower with *N*-carbamates[*24*] and it is known [*23*] that *N*-ethoxycarbonyl alanine can be converted to the corresponding acid chloride using oxalyl chloride/DMF without racemization. Furthermore, it is known that acetyl chloride reacts with aromatic aldehydes to form α-chlorobenzyl acetates in the presence of Lewis acids (*e g* $ZnCl_2$).[*25*] With this information at hand, our hypothesis was that the *in situ* generation of an acid chloride from a carbamate-protected amino acid followed by reaction with an aromatic aldehyde could afford an oxazolidinone directly.

Thus, treatment of (L)-*N*-*i*-butoxycarbonylalanine (**27**) in CH_2Cl_2 with catalytic DMF and 1 equivalent oxalyl chloride gives complete conversion to the corresponding acid chloride **28**. This was isolated as a yellow oil or used directly in solution. Addition of benzaldehyde followed by a catalytic amount of $SnCl_4$ in CH_2Cl_2 or solid $ZnCl_2$ leads to the formation of **29** in good yield as a thermodynamic mixture of *cis* and *trans* diastereomers in 85:15 ratio as outlined in Scheme 8. We then studied the formation of oxazolidinones from different carbamates and aldehydes. Given that epimerization at the C(2) position readily takes place upon exposure to a variety of acidic reagents such as $ZnCl_2$ in CH_2Cl_2 or *p*-toluenesulfonic acid in refluxing benzene, the reaction led to equilibrium compositions of *cis* and *trans* isomers, and this is shown in **Table I**.

Scheme 8

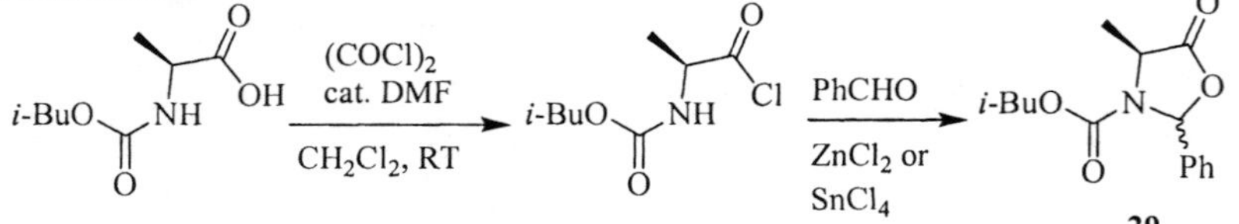

i-BuO, NH, O, OH → $(COCl)_2$, cat. DMF / CH_2Cl_2, RT → *i*-BuO, NH, O, Cl → PhCHO / $ZnCl_2$ or $SnCl_4$ → *i*-BuO, N, O, O, Ph

27 **28** **29**

Table I. Equilibrium ratios of different oxazolidinones

Entry	*R*	*Ar*	**Cis : Trans**	*Product*
1	Me	Ph	85 : 15	**29a**
2	Me	4`-biphenyl	85 : 15	**29b**
3	*i*-Bu	Ph	85 : 15	**29c**
4	*i*-Bu	4`-biphenyl	85 : 15	**29d**
5	*i*-Bu	9-anthranyl	55 : 45	**29e**

The preference for the *cis*-diastereomer is thought to arise from $A^{1,3}$-strain, with the nitrogen atom being pyramidalized and the *N*-acyl substituent pointing in the opposite direction from the substituents at C(2) and C(4), thus slightly lowering the free energy of the *cis*-isomer. In entry 5, the ratio of *cis* and *trans* is almost equal, and this is likely due to the steric bulk of the 9-anthranyl substituent leading to a very small energy difference between the two isomers. Since the subsequent alkylation requires a diastereomerically pure substrate, our goal was to try to find an oxazolidinone with crystalline properties. Ideally, in analogy with the imidazolidinone approach discussed above, we wanted to develop a crystallization-driven transformation, which would provide us with a pure diastereomer that could be isolated by simple filtration.

Our attempts to obtain a crystalline oxazolidinone were eventually successful with the (L)-*N*-*i*-butoxycarbonyl derivatives and 4`-biphenylcarboxaldehyde, leading to **29d**.[*26*] In solution, the typical 85:15 equilibrium ratio of *cis* and *trans* diastereomers is observed (**Table I**, entry 4). When the solvent was partially removed and the residue suspended in *tert*-butyl methyl ether, an off-white solid precipitated, which consists of pure *cis*-**29d** in 88% yield. Analysis of the mother liquors shows very minor amounts of *cis* and *trans* product in a 85:15 ratio, along with some unreacted carbamate and α,α-dichloro biphenyl. The presence of both *cis* and *trans* isomers in the mother liquor is clear evidence that a crystallization-induced asymmetric transformation is responsible for the high selectivity.

The structure of **29d** was confirmed by X-Ray crystallography and the *cis*-relationship of the 4-methyl and the 2-biphenyl substituents is readily verified in Figure 1.

Mechanistic considerations

The reaction of (L)-*N*-*i*-butoxycarbonyl-alanine and $(COCl)_2$ is rapid at room temperature and forms acid chloride **28** in virtually quantitative yield as determined by ^{1}H-NMR in CD_2Cl_2. When ^{13}C-labelled benzaldehyde (δ=192.2 ppm for $^{13}C=O$) is added to the acid chloride, a new signal is observed at 83.37 ppm by ^{13}C-NMR. Parallel to this, but at a slightly slower rate, new peaks grow

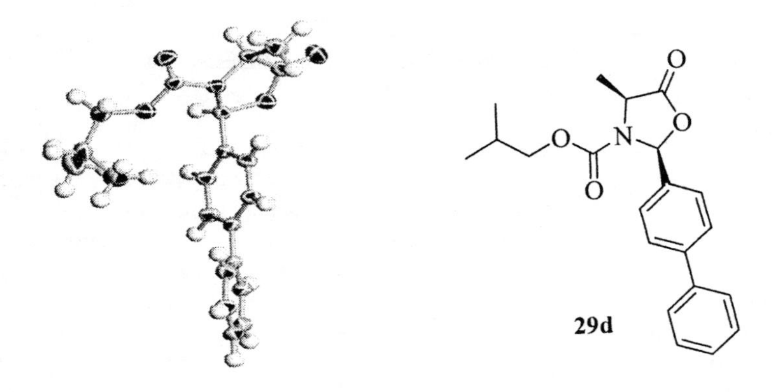

***Figure 1**. Single crystal X-ray structure (ORTEP representation) for oxazolidinone **29d**.*

Scheme 9

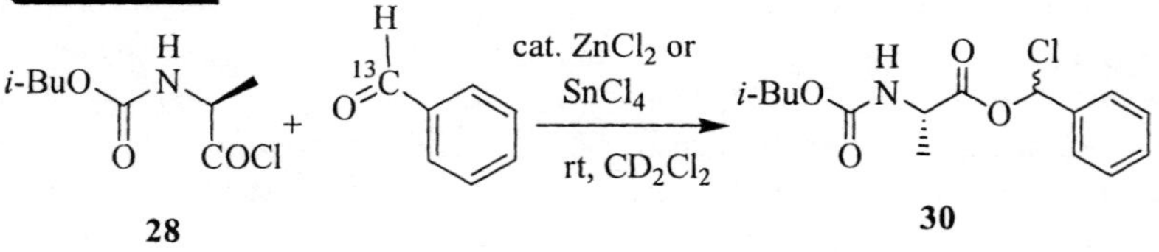

H
i-BuO N
O COCl
\+
H
^{13}C
O

cat. $ZnCl_2$ or $SnCl_4$

rt, CD_2Cl_2

H O Cl
i-BuO N O
O

28

30

in at 89.26 and 89.48 ppm. These peaks correspond to the signal for C(2) in *cis*- and *trans*-**29c**, respectively. By the end of the reaction, all of the unknown species observed at 83.37 ppm is consumed and the two diastereomers of **29c** are the two main products present. We propose that the unknown species with a signal at 83.37 corresponds to an intermediate α-chloro benzyl ester **30** as shown in Scheme 9.

The main support for intermediate **30** comes from correlation to the ^{13}C-NMR study of the analogous reaction with acetyl chloride (**31**) and ^{13}C-labeled benzaldehyde in the presence of catalytic $SnCl_4$ to give α-chlorobenzyl acetate (**32**), as laid out in Scheme 10. The signal for the benzylic carbon in **32** is readily observed at 82.86 ppm. This is very close to the signal at 83.37 ppm observed for the unknown intermediate (*vide supra*).

Scheme 10

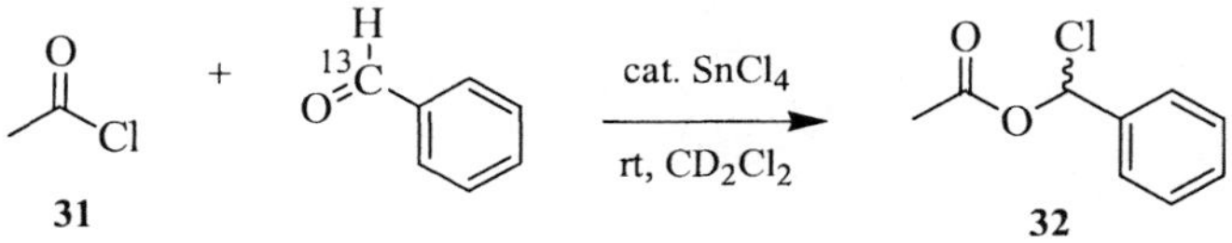

31 + ^{13}C-benzaldehyde → (cat. $SnCl_4$; rt, CD_2Cl_2) → **32**

Unfortunately, the isolation of **30** was not possible, because it underwent rapid cyclization to the oxazolidinone. Additional qualitative evidence for an intermediate in the cyclization comes from ReactIR experiments. When benzaldehyde is added to a CH_2Cl_2 solution of the acid chloride in the presence of catalytic $SnCl_4$, a new carbonyl resonance at 1800 cm^{-1} is observed, which is attributed to the oxazolidinone product. The benzaldehyde signal at 1702 cm^{-1} disappears at ca. 3 times faster rate than the appearance of the oxazolidinone signal, suggesting the formation of an intermediate which is transformed into the oxazolidinone product at a slightly slower rate. Scheme 11 shows the proposed mechanism for oxazolidinone formation.

Scheme 11

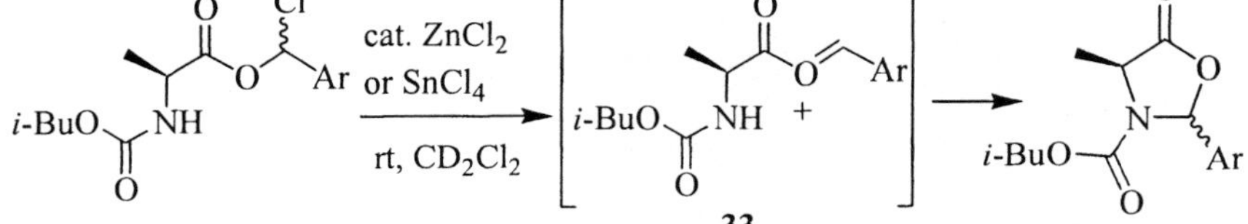

30 → (cat. $ZnCl_2$ or $SnCl_4$; rt, CD_2Cl_2) → [**33**] → **29**

Synthesis of BIRT 377

Alkylation of the enolate of **29d** is stereospecific and takes place from the face opposite the biphenyl moiety. It is highly diastereoselective and gives derivatives of 23 in >98% ee as determined by chiral HPLC after conversion to the methyl ester. In our procedure, LiHMDS (1.1 eq.) is added at -25 °C to a mixture of *p*-bromobenzyl bromide and 29d leading to 34 in virtually quantitative yield. The results are poorer when the enolate is first generated at -25 °C followed by addition of electrophile, indicating that the enolate itself is less stable at -25 °C.

Scheme 12

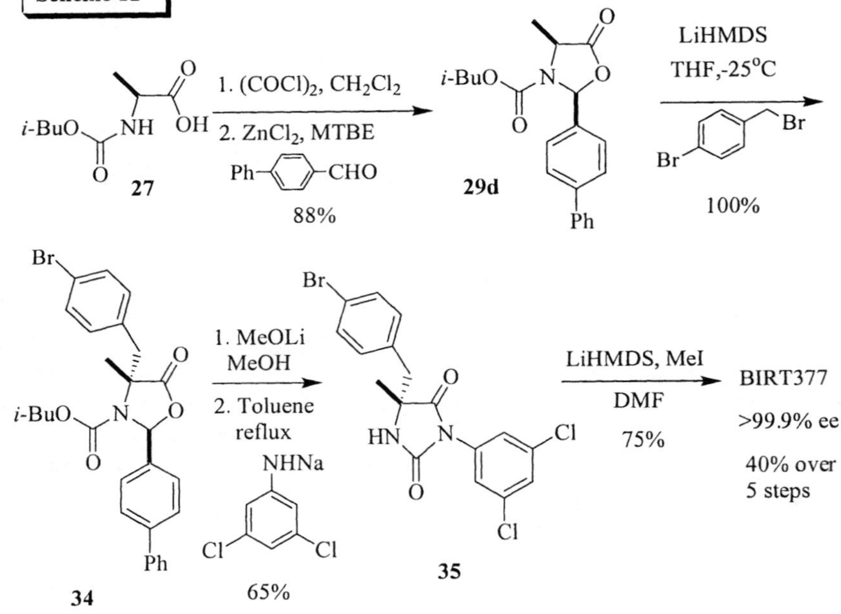

i-BuO NH OH O O **27** → 1. $(COCl)_2$, CH_2Cl_2; 2. $ZnCl_2$, MTBE, Ph–C$_6$H$_4$–CHO; 88% → *i*-BuO N O O O Ph **29d** → LiHMDS, THF,-25°C, Br–C$_6$H$_4$–CH$_2$Br; 100% →

Br *i*-BuO N O O O Ph **34** → 1. MeOLi, MeOH; 2. Toluene reflux, NHNa Cl Cl; 65% → Br HN N O O Cl Cl **35** → LiHMDS, MeI, DMF; 75% → BIRT377 >99.9% ee, 40% over 5 steps

The conversion of **34** into BIRT 377 is rather efficient and very atom-economical. Treatment with 1.1 eq. MeOLi in methanol at rt provides the methyl ester of the protected amino acid along with 4-phenyl benzaldehyde. Addition of aqueous sodium bisulfite precipitates the aldehyde as the bisulfite adduct which can be removed by filtration. The crude methyl ester is then refluxed in toluene with the sodium salt of 3,5-dichloroaniline to afford hydantoin **35**, which is then *N*-methylated using LiHMDS/MeI in DMF to provide the final product. This sequence provides BIRT 377 in excellent purity in 40% overall yield over 5 steps, starting in this case from readily available (*L*)-*N*-*i*-butoxycarbonylalanine.

Conclusion

In this chapter, we have shown how a discovery route can be quickly adapted into an "expedient" route, which is usually sufficient to produce kilogram amounts of a target compound, although in this case through a rather laborious process. We have then described two practical routes, both exploiting the principle of "self-regeneration of stereocenters" introduced by Seebach. The overall selectivity of this process was boosted in both cases by a novel and unusual crystallization-driven transformation. This, combined with a completely stereoselective alkylation, contributed to an overall 100% diastereoselective process, which is often a necessity in the development of practical chemistry, i.e. chemistry that does not rely on separative techniques like fractional crystallization or chromatography, two indicators of synthetic ineffectiveness.

Although both the above diastereoselective processes are practical and economical, the oxazolidinone route is more synthetically efficient, in that it completely does away with protecting groups (the carbamate carbonyl is finally utilized and incorporated in the final target), and utilizes the more economical (L)-alanine. Its drawback is the need to recycle 4-phenylbenzaldehyde, which is needed to achieve the desired crystallinity.

Overall, the imidazolidinone process achieves 6.8 % atom efficiency, consistent with the extensive use of blocking groups. This feature is not present in the oxazolidinone route, which has an atom economy of 48.5%, an indicator of a much higher synthetic efficiency.

References

1. Kelly, T.A.; Jeanfavre, D. D.; McNeil, D. W.; Woska, J.; Reilly, P. L.; Mainolfi, E. A.; Kishimoto, K. M.; Nabozny, G. H.; Zinter, R.; Bormann, B. J.; Rothlein, R. *J. Immunol.* **1999**, *163*, 5173.
2. Last-Barney, K.; Davidson, W.; Cardozo, M.; Frye, L. L.; Grygon, C. A.; Hopkins, J. L.; Jeanfavre, D. D.; Pav, S.; Qian, C.; Stevenson, J. M.; Tong, L.; Zindell, R.; Kelly, T. A. *J. Am. Chem. Soc.* **2001**, *123*, 5643.
3. Cativiela, C.; Diaz-de-Villegas., M. D. *Tetrahedron Asymmetry* **1998**, *9*, 3517.
4. Anderson, N.G., *Practical Process Research and Development.* Academic Press, San Diego, CA, **2000**.
5. Yee, C.; Blythe, T. A.; McNabb, T. J.; Walts, A. E. *J. Org. Chem.* **1992**, *57*, 3525.
6. Spero, D.M., Kapadia, S. R. *J. Org. Chem.* **1996**, *61*, 7398.
7. Yee, N. *Org. Lett.* **2000**, *2*, 2781.

8. Seebach, D.; Sting, A. R.; Hoffmann, M. *Angew. Chem. Int. Ed. Engl.* **1996**, *35*, 2708.
9. Naef, R.; Seebach, D. *Helv. Chim. Acta* **1985**, *68*, 135.
10. Seebach, D.; Aebi, J. D.; Naef, R.; Weber, T. *Helv. Chim. Acta* **1985**, *68*, 144.
11. Aebi, J.D.; Seebach, D. *Helv. Chim. Acta* **1985**, *68*, 1507.
12. Seebach, D.; Dziadulewicz, E.; Behrendt, L.; Cantoreggi, S.; Fitzi, R. *Liebigs. Ann. Chem.* **1989**, 1215.
13. Grozinger, K.G.; Kriwacki, R. W.; Leonard, S. F.; Pitner, T. P. *J. Org. Chem.* **1993**, *58*, 709.
14. Seebach, D.; Gees, T.; Schuler, F. *Liebigs Ann. Chem.* **1993**, 785.
15. Kazmierski, W.M.; Urbanczyk-Lipkowska, Z.; Hruby, V. J. *J. Org. Chem.* **1994**, *59*, 1789.
16. Studer, A.; Seebach, D. *Liebigs Ann. Chem.* **1995**, 217.
17. Ma, D.; Tian, H. *J. Chem. Soc. Perkin Trans. 1* **1997**, 3493.
18. Jaun, B.; Tanaka, M.; Seiler, P.; Kuhnle, F. N. M.; Braun, C.; Seebach, D. *Liebigs Ann. Chem.* **1997**, 1697.
19. Damhaut, P.; Lemaire, C.; Pienevaux, A.; Birhaye, C.; Christiaens, L.; Comar, D. *Tetrahedron* **1997**, *53*, 5785.
20. Frutos, R.P.; Stehle, S.; Nummy, L.; Yee, N. *Tetrahedron: Asymmetry* **2001**, *12*, 101.
21. Kapadia, S.R.; Spero, D. M.; Eriksson, M. *J. Org Chem.* **2001**, 1903.
22. Micheel, F.; Meckstroth, W. *Chem. Ber.* **1959**, ***92,*** 1675.
23. Buckley III, T.F.; Rapoport, H. *J. Am. Chem. Soc.* **1981**, *103*, 6157.
24. Jones, J.H.; Witty, M. J. *J. Chem. Soc. Perkin Trans. 1* **1979**, 3203.
25. Neuenschwander, M.; Bigler, P.; Christen, K.; Iseli, R.; Kyburz, R.; Mühle, H. *Helv. Chim. Acta* **1978**, *61(6)*, 2047.
26. Napolitano, E.; Farina, V. *Tetrahedron Lett.* **2001**, *42*, 3231.

Chapter 3

Synthetic, Structural, and Mechanistic Issues in the Development of a Subtype Selective GABA Partial Agonist Hypnotic Agent

John A. Ragan[1,*], Jerry A. Murry[1,2], Michael J. Castaldi[1], Alyson K. Conrad[1], Paul D. Hill[1], Brian P. Jones[1], Narasimhan Kasthurikrishnan[1], Bryan Li[1], Teresa W. Makowski[1], Ruth McDermott[1], Barb J. Sitter[1], Timothy D. White[1], and Gregory R. Young[1]

[1]Chemical Research and Development, Pfizer Global Research and Development, Eastern Point Road, Groton, CT 06340
[2]Current address: Chemical Process Development, Merck and Company, Rahway, NJ 07065
[*]Corresponding author: telephone: 860–441–6334, fax: 860–441–5445, email: john.a.ragan@groton.pfizer.com

A variety of synthetic studies focused on clinical CNS candidate **1** are described. The original medicinal chemistry route (Scheme 1) is described, and issues which precluded its scale-up are discussed. An Ullman route to 2-fluoro-4-methoxyaniline (Scheme 4) was developed to avoid a non-selective nitration reaction. The first GMP bulk campaign utilized a ring expansion strategy *via* a dichloroketene [2+2] cycloaddition (Scheme 6) to prepare cycloheptane-1,3-dione. While effective on laboratory scale, several issues arose upon scale-up; the mechanistic basis for these issues was determined to be competition between desilylation and dechlorination of dichlorocyclobutanone **22** (Scheme 11). These issues led us to develop a third synthesis of **1**, in which cycloheptane-1,3-dione is avoided. Two variants of this Friedel-Crafts strategy are described (Schemes 13 and 14).

Introduction

N-(2-Fluoro-4-methoxyphenyl)-4-oxo-1,4,5,6,7,8-hexahydrocyclohepta [b]pyrrole-3-carboxamide (**1**) is a sub-type selective partial agonist and modulator of the GABA receptor complex, (*3*) and was recently of interest for clinical evaluation as an agent for the treatment of sleep disorders such as insomnia. The pharmacological selectivity and rapid clearance of partial agonists such as **1** are anticipated to provide advantages over existing therapies (e.g. Sonata®, Ambien®) in terms of potential for addiction and lack of residual next day side effects. At the time of its identification as a drug candidate, multigram quantities of **1** had been prepared. The studies required for advancement of this compound into the clinic (e.g. regulatory toxicology, formulation development, manufacture of clinical supplies) required significantly larger quantities of material prepared under more controlled, cGMP (current Good Manufacturing Practice) conditions, as well as more stringent analytical standards and characterization. As is frequently the case, the synthesis utilized in Discovery to prepare the initial supplies of **1** (0.1-50 g) was not suitable for scale-up in the Pilot Plant. This chapter will describe the synthetic studies pursued to facilitate preparation of kilogram quantities of **1**, as well as several related chemical issues (synthetic, mechanistic, and structural) which impacted the clinical development of **1**.

Figure 1

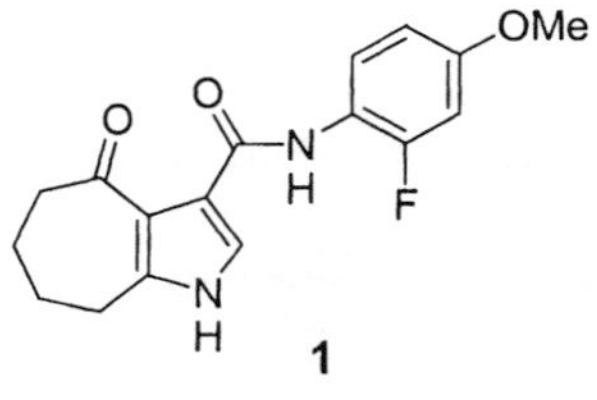

1

Discovery Synthesis

The first synthesis of **1** is outlined in Scheme 1. This synthesis was more than adequate for the goals it served: rapid preparation of the compound for initial *in vitro* potency experiments, followed by modest scale up (10-50 g) to support initial *in vivo* experiments and further pharmacological and pharmacokinetic characterization.

Scheme 1

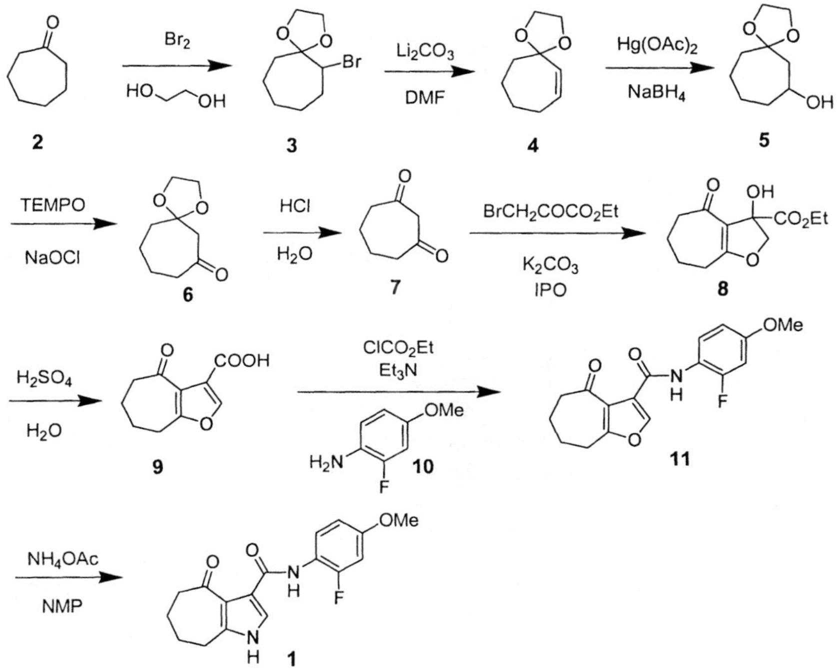

Cycloheptane-1,3-dione (**7**) is a key intermediate in the original synthesis. Several literature routes to this diketone are known, (*4*) one of which is shown in Scheme 1.(*4f*) For the preparation of decagrams of cycloheptane-1,3-dione, this route is perfectly acceptable. However, for scale-up to multi-kilograms under cGMP conditions in a pilot plant, we had serious reservations about the oxymercuration of olefin **4**. Use of stoichiometric $Hg(OAc)_2$, which forms an organomercurial intermediate (prior to demercuration with $NaBH_4$), poses several problems including worker safety, tank contamination, and contamination of final drug substance; analysis for *ppm* levels of mercury was anticipated, which would complicate and slow down the analytical release of drug substance. Moreover, tank contamination would be a serious issue; based on previous experience, we felt that the only way forward with this reaction would be to run it in 22 L glassware, which could be disposed of after processing. Although this is an option under certain scenarios, we were seeking to prepare 10-20 kg quantities of diketone **7**. The solvent volumes for this reaction were such that the throughput in 22 L glassware would be severely limiting, even for a first campaign.

Scheme 2

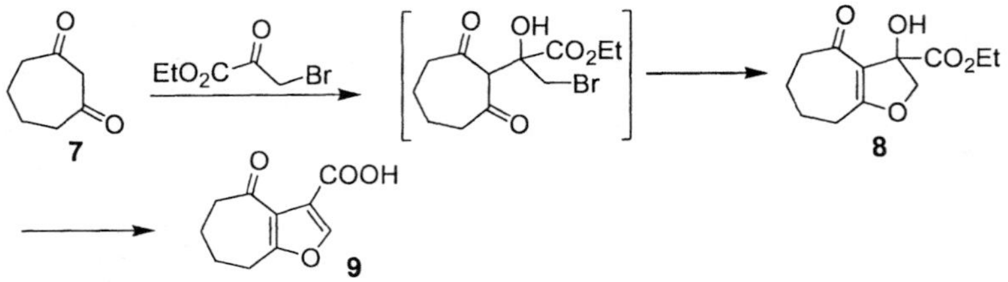

The furan annulation sequence (**7** → **8** → **9**) of the original synthesis works quite well: with reasonably pure diketone, the yield for this sequence is very good (80-90%), and furan acid **9** is a well-behaved solid which can be easily isolated by recrystallization (Scheme 2). The intermediate tertiary alcohol **8** can be isolated as an oil, but it was found to be operationally simpler to take the crude product into the acidic hydrolysis and isolate **9** as a solid. This type of furan annulation of a 1,3-diketone is well precedented in the literature for cyclic diketones such as cyclohexane-1,3-dione (*5*) and dimedone (5,5-dimethyl-1,3-cyclohexanedione.(*6*) Based on these observations, our initial strategy for avoiding the oxymercuration was to seek an alternative route to cycloheptane-1,3-dione, and utilize the downstream chemistry described in Scheme 1. We first turned our attention to the synthesis of the requisite aniline, 2-fluoro-4-methoxyaniline (**10**).

Preparation of the Aromatic Side Chain: Ullman Methoxylation in the Presence of a Protected Aniline

Although known, the literature synthesis of 2-fluoro-4-methoxyaniline (*7*) outlined in Scheme 3 presented scale-up problems. The synthesis involves a non-selective nitration of 3-fluorophenol. Separation of the desired regioisomer, methylation, and nitro reduction provides the target aniline (**10**). Finding a more practical synthesis of this aniline was the first issue addressed on this project.

Scheme 3

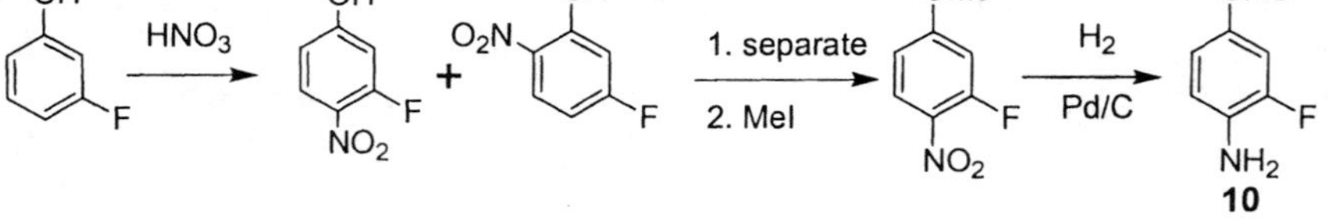

The commercial availability of 2,4-difluoronitrobenzene (**12**) motivated us to investigate methoxide displacement of one of the fluorines, hoping to find conditions which favored displacement of the fluoro group *para* to the nitro substituent to provide **13** (Nu = OMe, Table 1). As summarized in Table 1, displacement with NaOMe favored displacement of the *ortho* fluorine, to provide the undesired regioisomer **14** (entries 1-3, Nu = OMe). The ratio of products varied with choice of solvent, from 1:2 (MeOH, DMF) to 1:8 (THF). Displacement with hydroxide was also studied (entries 4-7), but uniformly favored formation of the undesired *ortho* substitution product **14** (Nu = OH).

Table 1: Nucleophilic Substitution of 2,4-difluoronitrobenzene

Entry	*Nu*	*Conditions*	*13* (*% GC yield*)	*14* (*% GC yield*)
1	OMe	NaOMe, MeOH	33	67
2	OMe	NaOMe, DMF	33	67
3	OMe	NaOMe, THF	11	89
4	OH	0.6 N NaOH	14	86
5	OH	6 N NaOH, DMSO	14	86
6	OH	0.6 N LiOH	17	83
7	OH	1 N NaOH, IPE, Bu_4NHSO_4	17	83

These results are not inconsistent with literature precedents for the S_NAr reaction of 2,4-difluoronitrobenzene. There are several examples of displacement with oxygen, (*8*) nitrogen, (*9*) and carbon (malonate) (*10*) nucleophiles. With oxygen nucleophiles, regioselectivity ranges from non-selective (1:1) to modest *ortho* selectivity (2-3:1). With nitrogen nucleophiles the regioselectivity varies more widely, from *para*-selective (*9a*) to non-selective, (*9b*) to *ortho*-selective.(*9c,d*) With stabilized carbon nucleophiles (malonate), there are reports of both *para* (*10a*) and *ortho* (*10b*) selectivity.

We next turned our attention to utilization of 2-fluoro-4-iodoaniline (**15**), hoping to displace the iodide with methoxide under Ullman-type conditions.(*11*) Literature precedent for effecting this type of coupling on the free aniline suggested that it would be less efficient than for related aryliodides, (*12*) and indeed, direct treatment of 2-fluoro-4-iodoaniline with NaOMe and CuCl in DMF generated none of the desired product. We next examined several blocking groups for the aniline.(*13*) Groups containing an acidic NH such as acetamide ($ArNHCOCH_3$) or methyl carbamate ($ArNHCO_2Me$) were unsuccessful: no product was observed, and after prolonged reaction times, cleavage of the blocking group became problematic. The phthalate derivative was found to be unstable to the basic reaction conditions. Turning to a base-stable blocking group, we prepared the benzophenone imine ($ArN{=}CPh_2$). While stable to the reaction conditions, this substrate led to a 1:1 mixture of methoxide displacement at the fluorine and iodine, apparently due to the anion stabilizing properties of the imine. Success was realized with the 2,5-dimethylpyrrole derivative **16** (Scheme 4). This blocking group lacks an acidic NH, is stable to the basic reaction conditions, and is not sufficiently activating to induce displacement of the fluorine. Scheme 4 shows its synthesis, Ullman coupling, and deprotection; this sequence was scaled successfully to provide 10 kg quantities of the requisite aniline **10**.(*14*)

Scheme 4

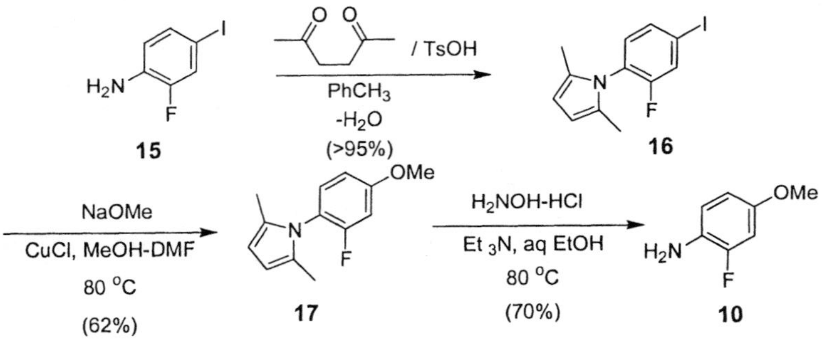

15 → (/ TsOH, $PhCH_3$, $-H_2O$, (>95%)) → **16** → (NaOMe, CuCl, MeOH-DMF, 80 °C, (62%)) → **17** → (H_2NOH-HCl, Et_3N, aq EtOH, 80 °C, (70%)) → **10**

Enabling the First Campaign: A Ketene [2+2] Route to Cycloheptane-1,3-dione

With the requisite aniline in hand, we turned our efforts to developing a scaleable route to furan acid **9** (Scheme 1). As discussed previously, concerns with the oxymercuration reaction motivated us to seek an alternative route to cycloheptane-1,3-dione, which we knew was a viable precursor to the requisite furan acid (Scheme 2). Of the known syntheses of cycloheptane-1,3-dione, (*4*) one which intrigued us was Noyori's report (*4d*) of a palladium-mediated isomerization of 2,3-epoxy-1-cycloheptanone (**20**) to the 1,3-diketone (Scheme 5). The requisite epoxide is available from cyclohept-2-en-1-one (**19**); although prohibitively expensive for multi-kilo supplies ($179 / 10 g), (*15*) this was a convenient starting material for laboratory scale investigations. Alternatively, the enone is available in two steps from the more readily available cycloheptanone ($67 / 250 g) (*15*) *via* α-bromination (*16*) and elimination.(*17*) Noyori's conditions (4-8 mol% $Pd(Ph_3P)_4$ and 1,2-bis(diphenylphosphino)ethane (dpe)) converted epoxide **20** to the desired diketone (**7**) in 12-16 h. We found that using BINAP in place of dpe allowed the catalyst load to be reduced to 2-3 mol% without adversely affecting the rate of rearrangement. The initial experiments were conducted with non-racemic BINAP; as anticipated, both (*S*)- and (*R*)-enantiomers worked equally well (the substrate epoxide is racemic, and the product is achiral). Surprisingly, when racemic BINAP was utilized, incomplete conversion was observed (<20%), with significant quantities of an M+2H reduction product observed by mass spec (presumably from hydrogenolysis of one of the two epoxide C-O bonds). We suspected that a different purity profile for the racemic vs. single enantiomer BINAP was responsible for this (possibly due to differing crystal forms and thus recrystallization properties). However, when an "artificial racemate" was utilized with 1.5 mol% each of (*S*)- and (*R*)-enantiomers, the reaction again failed. Thus, there appears to be a true difference in the behavior of racemic BINAP from the single enantiomer in this reaction, a result which has yet to be adequately explained (one possibility is different catalytic activities of a Pd(*R*-BINAP)(*S*-BINAP) species from the , $Pd(R\text{-BINAP})_2$ or $Pd(S\text{-BINAP})_2$) species.

Scheme 5

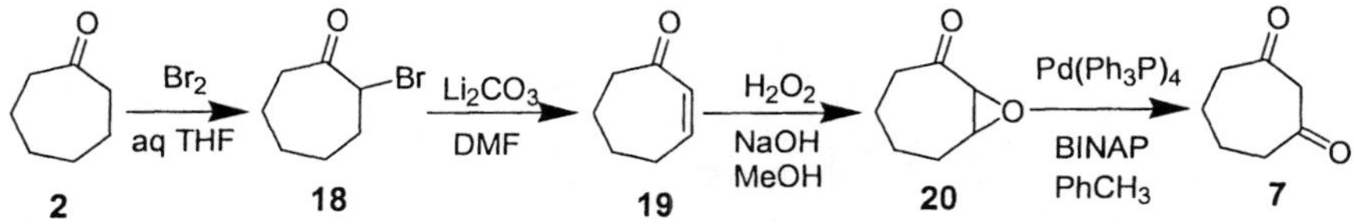

Br_2 / aq THF; Li_2CO_3 / DMF; H_2O_2 / NaOH / MeOH; $Pd(Ph_3P)_4$ / BINAP / $PhCH_3$

2 → **18** → **19** → **20** → **7**

The epoxide rearrangement route was ultimately deemed impractical for scale-up due to the cost of catalyst and the requirement for a minimum of 2-3 mol% (which translates to 28 and 15 weight% of $Pd(Ph_3P)_4$ and BINAP, respectively). Another route investigated was the palladium-mediated oxidation of cyclohept-2-en-1-one (**19**) with *tert*-BuOOH and $PdCl_2$. Consistent with literature precedent, (*18*) this oxidation was regioselective for the desired 1,3-diketone. Unfortunately, a *tert*-butyl--containing impurity contaminated the product and could not be removed by non-chromatographic methods. This impurity adversely affected the furan annulation sequence, and this approach was abandoned in favor a more efficient ring expansion method described below.

A shorter, more practical synthesis of diketone **7** was realized upon consideration of a two carbon ring expansion from cyclopentanone (Scheme 6).(*19*) The known (*20*) [2+2] cycloadduct **22** (prepared from dichloroketene and TMS enol ether **21**) contains the requisite seven carbon atoms, and possesses the desired 1,3-oxygen substitution pattern. Conversion of **22** into cycloheptane-1,3-dione requires reduction of the two chlorine atoms, desilylation of the tertiary alcohol, and retro-aldol fragmentation of the resulting β-hydroxyketone (the ordering of these events will turn out to be critical, *vide infra*). Indeed, this strategy has been demonstrated by Pak et al., (*21*) who reported a two-step conversion of 3-trimethylsiloxy-2,2-dichlorocyclobutanones into 1,3-diketones *via* Bu_3SnH-mediated dechlorination to the 3-trimethylsiloxycyclobutanone, followed by Bu_4NF-induced desilylation and ring opening to the diketone.(*22*) While this work provided excellent precedent for this strategy, Bu_3SnH was clearly unacceptable for our purposes for the same reasons outlined with the original organomercurial intermediate (Scheme 1).

Scheme 6

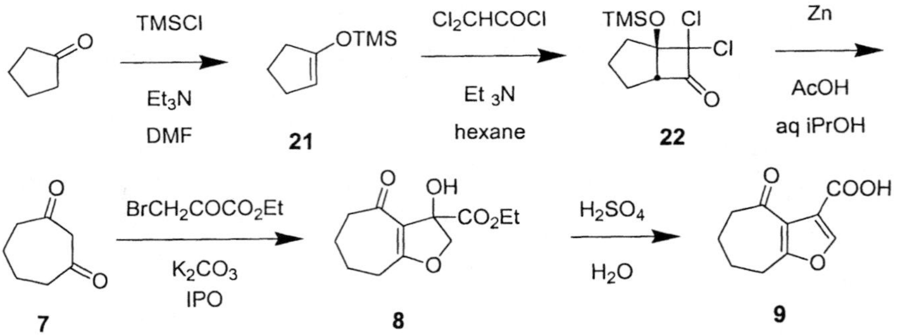

TMSCl, Et_3N, DMF → **21** (OTMS); $Cl_2CHCOCl$, Et_3N, hexane → **22** (TMSO, Cl, Cl, O); Zn, AcOH, aq iPrOH → **7**; $BrCH_2COCO_2Et$, K_2CO_3, IPO → **8** (OH, CO_2Et); H_2SO_4, H_2O → **9** (COOH)

Zinc-acetic acid is a well precedented reagent for reduction of 2,2-dichloroketones.(*23*) While numerous examples of this type of reduction are found in the literature, (*24*) at the time we were pursuing this work there were no examples with 3-trimethylsiloxy-2,2-dichlorocylobutanones (e.g. **22**). Thus, we were pleased to find that treatment of dichlorocyclobutanone **22** with zinc (either powdered or –30+100 mesh granules) and acetic acid in aqueous isopropanol cleanly provided diketone **7** on laboratory scale. The overall yield of crude diketone from cyclopentanone is 58-64%. This material is a dark orange oil, which is 80-90% pure as indicated by ^{1}H NMR analysis. The diketone can be purified by distillation to remove the orange color, however, this also causes significant degradation of the diketone. The isolated yield following distillation is 38% from cyclopentanone (ca. 60-65% recovery of the crude mass). The ^{1}H NMR spectra before and after distillation are virtually indistinguishable, so it was preferred to utilize the crude diketone directly in the furan annulation sequence and isolate furan acid **9** as a crystalline solid (use of distilled diketone did not improve the overall yield from cyclopentanone). On laboratory scale (19 g of cyclopentanone), the conversion of cyclopentanone to furan acid **9** proceeds in 33% yield over the five-step sequence (an average of 80% per step). However, upon scale-up, the overall yield suffered considerably, as will be discussed later.

On laboratory scale runs of the Zn/AcOH reduction (up to 150 g of the dichlorocyclobutanone intermediate), moderate exotherms were observed, particularly at the beginning of the AcOH addition. Concerns regarding the heat flow associated with multi-kilogram reductions led us to examine the reaction in a Mettler RC1 reaction calorimeter.(*17*) This was run on 50 g of dichlorocyclobutanone **22** in 270 mL of 50% aqueous isopropanol, with dropwise addition of 100 mL of 50% aqueous acetic acid over a period of 4 h. These studies showed an exotherm of 500 kJ/mol (120 kcal/mol), which translates to a maximum adiabatic temperature rise of 65-70 °C. Thus, the "worst case scenario" in terms of safety, i.e., if the AcOH were added in a single portion with no external cooling, would lead to a temperature rise from 25 °C to 90-95 °C. From a safety perspective this was viewed as acceptable, as the boiling point of the reaction mixture is above this temperature range. This reaction sequence was scaled to 100 kg of **22** without safety incident, although the chemical outcome was disappointing (*vide infra*).

End Game Chemistry: Identification and Minimization of an M+NH Impurity

The conversion of furan acid **9** to the drug candidate **1** is straightforward (Scheme 7). The aniline side chain is installed *via* the mixed anhydride, and the activated furan is converted to the pyrrole by treatment with NH_4OAc in hot *N*-methylpyrrolidinone (NMP). The latter reaction benefits from a particularly efficient work-up: addition of water to the warm NMP solution induces crystallization of the product, which is filtered directly from the reaction mixture. HPLC purities of >98% are typical from this reaction, and the product can be recrystallized if further purification is needed. We struggled with one impurity, which was present at levels between 0.5-1.0% in the crude product; recrystallization reduced the level to 0.3-0.5%. We initially did not know the identity of this impurity, other than by mass spectral analysis, which indicated a molecular ion of M+NH (**1** has formula $C_{17}H_{17}FN_2O_3$, the impurity is

$C_{17}H_{18}FN_3O_3$ by exact mass). Although the impurity was qualified in both 30 day and genetic toxicology studies, we knew that at some point its exact identity would be needed, so several hypotheses were proposed and experiments run to ascertain its structure.

Scheme 7

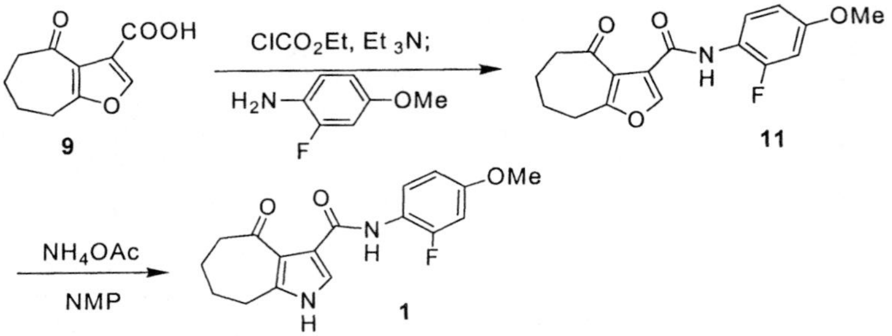

Some of the proposed structures, consistent with the mass spectral fragmentation data, are shown in Scheme 8. We reasoned that structures such as **23** or **24** might arise from adventitious oxidation during the reaction, which could generate hydrazine. We purposely ran the final step with hydrazine-acetic acid in place of NH_4OAc, and observed a new product with the anticipated M+NH mass spectrum; however, this material did not co-elute with our impurity. We also considered structures such as **25** or **26**, which could arise from a Beckman rearrangement of oxime **27** (formed from hydroxylamine, present either as an impurity in the reagent, or formed by adventitious oxidation during the reaction). Oxime **27** was prepared, and upon exposure to Beckman rearrangement conditions (AcOH, reflux), generated two new products with the expected mass spectrum. Again, neither product co-eluted with our impurity.

Scheme 8

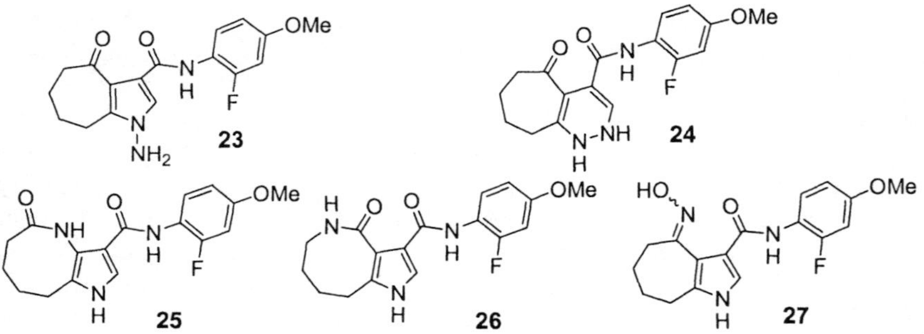

Success was realized with the proposal of structure **29** (Scheme 9). This material was synthesized by treatment of **1** with *para*-nitrobenzene diazonium chloride (from diazotization of the *para*-nitroaniline), followed by reduction of the diazo compound **28** with $SnCl_2$ in AcOH.(*25*) This 2-aminopyrrole co-eluted with the observed impurity, and displayed identical mass spec fragmentation and ^1H NMR spectra, confirming its structure as **29**.

Scheme 9

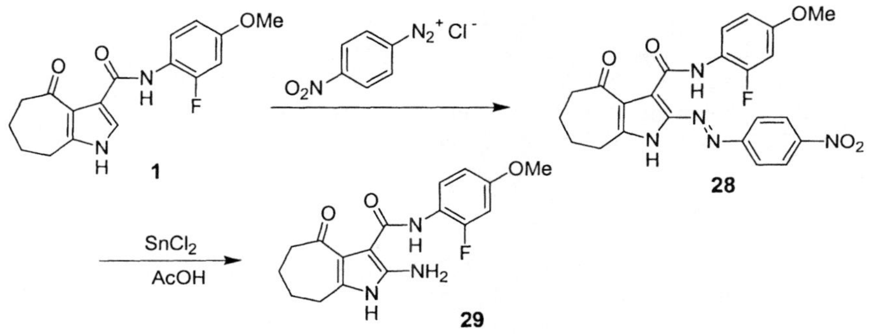

OMe O O N H F N H 1 O2N N2+ Cl- SnCl2 AcOH NH2 29 28 NO2

Compound **29** presumably arises from adventitious oxidation of one of the intermediates during the formation of **1**; one possible mechanism is shown in Scheme 10. Purging the reaction vessel with nitrogen prior to initiation of heating was found to minimize the level of this impurity.

Scheme 10

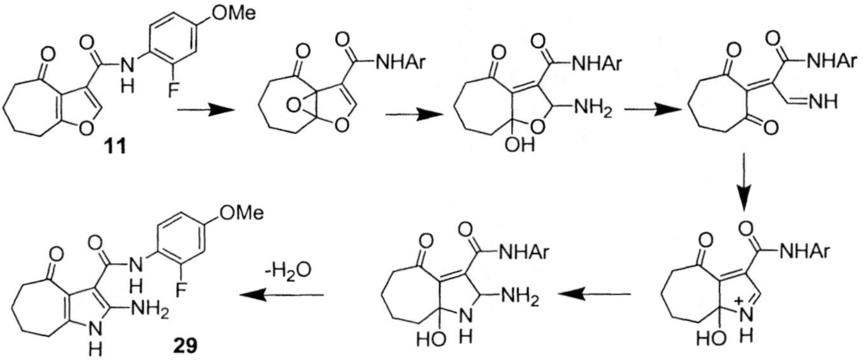

OMe O O N H F 11 NHAr NH2 OH NH -H2O HO H 29

An Improved Synthesis of the Keto-Furan: An Efficient Friedel-Crafts Cyclization

As described earlier, the dichloroketene route to cycloheptane-1,3-dione and thence furan acid **9** worked well on laboratory scale (Scheme 6). Upon scale-up in the pilot plant, however, the yield dropped dramatically: starting from 28.5 kg of cyclopentanone, 3.35 kg of **9** was obtained, representing an overall yield of just 5%. Moreover, isolation of solid **9** required a tedious series of silica gel filtrations, which did not bode well for further scale-up. The major contaminant in the crude furan acid was identified as 6-oxoheptanoic acid (**35**), which was traced back to 2-acetylcyclopentanone (**34**) formed during the Zn/AcOH reduction (Scheme 11). This was confirmed by subjection of 2-acetylcyclopentanone to the furan annulation sequence (ethyl bromopyruvate, K_2CO_3, *i*-PrOH; aq H_2SO_4), which cleanly formed 6-oxoheptanoic acid *via* a retro-Claisen reaction. These byproducts had been present at significantly lower levels in the laboratory scale runs (ca. 10%, vs. 40-45% upon scale-up).

An investigation of the formation of 2-acetylcyclopentanone determined that the level of this impurity was highly temperature dependent. When the reaction is kept at or below 0 °C, a minimal amount (5-10%) of this impurity is formed. However, when the reaction is run at 40-50 °C, significantly higher levels (40-50%) of the byproduct are seen. Our reasoning for this, as outlined in Scheme 11, is the temperature dependence of two competing processes: hydrolysis of the silyl ether versus chloro reduction. If both chlorines are reduced while the silyl ether is still intact (e.g. **30**), then hydrolysis and retro-aldol are selective for the seven-membered diketone (path A). (GC/MS monitoring of laboratory scale runs indicated that this was the predominant pathway: loss of each chlorine could be observed in the mass spectrum prior to hydrolysis of the silyl ether). However, if the silyl ether is cleaved while one (or both) chlorines are still intact (e.g. **32**), then hydrolysis and retro-aldol cleavage favors formation of cyclopentanone **33** (path B), (*20*) presumably due to the increased acidity of the dichloromethyl ketone. We tested this hypothesis by subjecting dichlorocyclobutanone **22** to aqueous acid in the absence of zinc; as expected, clean formation of **33** was observed by GC/MS and ^{1}H NMR.(*24*) Treatment of this material with Zn/AcOH generated acetylcyclopentanone (**34**).

Heat transfer processes can differ significantly based on reaction volume, since the heat transfer capacity of a given reaction vessel is proportional to its surface area (m^2), whereas reaction mass is proportional to volume (m^3). Given

Scheme 11

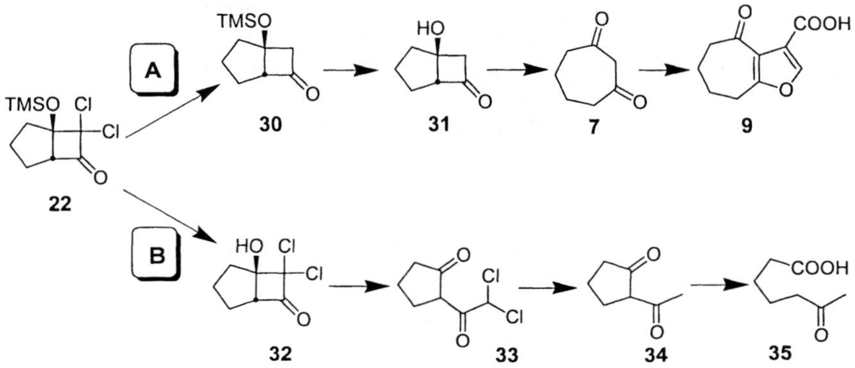

the strong temperature dependence for byproduct formation in this reaction, we felt that this problem was likely to be faced again with subsequent increases in scale (the increased level of byproduct in the pilot plant run was attributed to local heating during the addition of AcOH). The problem is exacerbated by the fact that none of the intermediates in Scheme 6 are solids prior to furan acid **9**; thus, the only clean-up opportunities prior to the end of the sequence are chromatography or distillation. We felt that a more robust synthesis would be required to support subsequent campaigns.

In seeking a new route to furan acid **9**, we felt that it would be advantageous to avoid cycloheptane-1,3-dione as an intermediate. The reasons for this were varied, but included:

Modest stability of the diketone. Cycloheptane-1,3-dione is a colorless oil. We found that crude samples of this material tended to discolor upon storage above 0 °C. This color could be removed by distillation or filtration through silica gel, but with associated loss of material (distillation was particularly inefficient, 60-65% recoveries were typical). This is in marked contrast to cyclohexane-1,3-dione, which is a stable, crystalline, white solid. Interestingly, the ^{1}H NMR spectra of these two materials in $CDCl_3$ are also quite different: the 6-membered diketone is completely enolized, whereas the 7-membered diketone exists exclusively in the diketo form. Transannular interactions or some other form of ring strain are presumably responsible for this difference in enolization propensity, and this correlates with the greater reactivity/instability of the 7-

membered diketone. House has reported similar observations with 2-phenylcycloheptane-1,3-dione.(*26*)

Limited options for purification of the diketone. As an oil, recrystallization was not available as a purification option. Chromatography and distillation were considered less attractive options. We spent considerable time trying to identify derivatives which would be crystalline, and could then be reverted back to the diketone, or incorporated directly into the furan annulation sequence (e.g. tosylhydrazones, bisulfite adducts, ketalization). These efforts were not fruitful.

Few options to significantly improve the current synthesis. The three step synthesis of cycloheptane-1,3-dione outlined in Scheme 6 is quite short, and we felt it was unlikely that we would be able to significantly shorten this sequence, or add crystalline intermediates. To the latter end, the dibromo and tert-butyldimethylsilyl analogs of **22** were prepared, but were not crystalline, and offered no significant advantage over the dichloro/TMS ether **22**.

Considering alternative routes to furan acid **9**, a variety of strategies were identified and pursued. These have been described elsewhere: they included a Dieckmann cyclization, one carbon ring expansion strategies from the 6-membered furan acid, and an intramolecular acetylene-furan Diels-Alder approach. (17) However, none of these provided a significant improvement to the route described in Scheme 6.

The most promising route arose from a retrosynthetic analysis which incorporated two Friedel-Crafts reactions (Scheme 12). The first disconnection involves formation of a 7-membered ring, which has recent precedent in Friedel-Crafts chemistry of furans.(*27*) Addition of a ketone (in a retrosynthetic sense) then generates a second Friedel-Crafts disconnection, and a well-precedented starting material derived from 3-carboxyfuran.(*28*)

Scheme 12

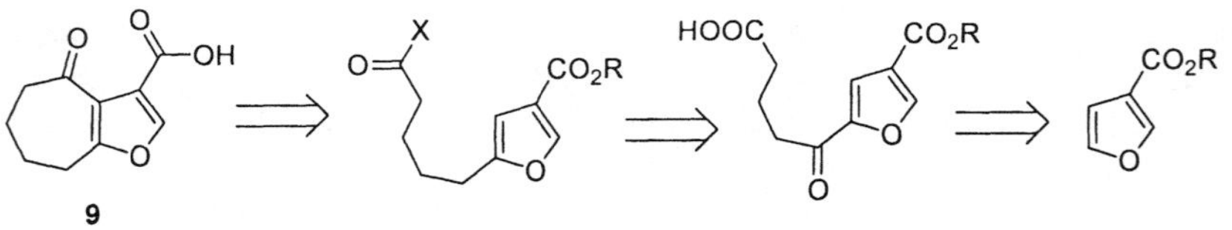

Two variants of this basic strategy were developed: the first route is shown in Scheme 13. Starting with ethyl 3-furancarboxylate, acylation with monomethylglutarate was effected with trifluoroacetic anhydride (TFAA) and phosphoric acid in acetonitrile to provide keto-furan **37** in 64% yield. Reduction of the ketone to **38** was achieved by treatment with Et_3SiH and $BF_3{\cdot}Et_2O$ (82%

yield on 5 g scale). Cyclization of diacid **38** was then effected by treatment with TFAA and $SnCl_4$, but this reaction was capricious and frequently failed to reach completion.

Scheme 13

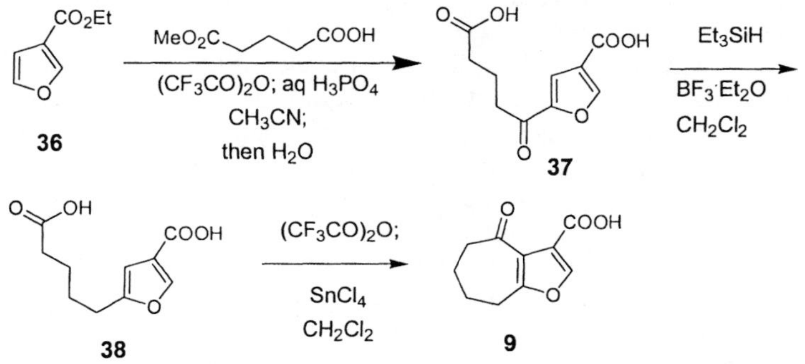

CO_2Et MeO_2C COOH ($(CF_3CO)_2O$; aq H_3PO_4, CH_3CN; then H_2O) OH COOH Et_3SiH ($BF_3 \cdot Et_2O$, CH_2Cl_2)

36 → 37

OH COOH ($(CF_3CO)_2O$; $SnCl_4$, CH_2Cl_2) COOH

38 → 9

The inefficiency of the final Friedel-Crafts cyclization was attributed to the presence of two non-equivalent carboxylic acids in **38**; activation of the carboxylate adjacent to the furan might lead to intermolecular oligomerization processes. We reasoned that if only the desired position was available for activation, a more efficient cyclization could be realized. This strategy required a method for differentiation of the two carboxylic acids, which was achieved as shown in Scheme 14.

Scheme 14

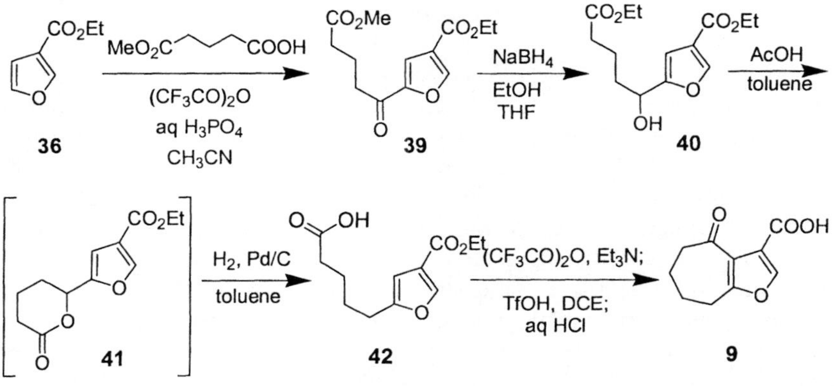

CO_2Et MeO_2C COOH ($(CF_3CO)_2O$, aq H_3PO_4, CH_3CN) CO_2Me CO_2Et ($NaBH_4$; EtOH, THF) CO_2Et CO_2Et OH (AcOH, toluene)

36 → 39 → 40

CO_2Et (H_2, Pd/C; toluene) OH CO_2Et ($(CF_3CO)_2O$, Et_3N; TfOH, DCE; aq HCl) COOH

[41] → 42 → 9

The initial Friedel-Crafts was run as before, but without a subsequent hydrolysis to isolate the bis-ester **39**. Reduction of the ketone with $NaBH_4$ provided alcohol **40**, which cylized to the corresponding δ-lactone (**41**) upon treatment with AcOH in toluene. Hydrogenolysis of the pseudobenzylic C-O bond then provided acid **42**, in which the desired carboxylate was differentiated from the ethyl ester. As anticipated, cyclization of this substrate was more efficient. Several combinations of acyl activations (trifluoroacetate mixed anhydride, acid chloride), Lewis acids ($AlCl_3$, $BF_3 \cdot Et_2O$, $SnCl_4$, H_3PO_4, CF_3SO_3H), and solvents (dichloromethane, CH_3CN, nitrobenzene, dichloroethane) were investigated. The major side products appeared to be higher molecular weight oligomers, from intermolecular Friedel-Crafts processes which competed with the desired cyclization. Minimization of these processes was complicated by the fact that the higher oligomers were generally not observable by our HPLC method, and revealed themselves upon work-up in the form of tarring. Thus, an experiment would seem to have proceeded cleanly to the desired product based on its HPLC profile, but upon workup and purification, the isolated yield of **9** would be modest. Inverse addition or high dilution conditions were found to reduce the formation of these oligomers.

Optimal conditions were found to be formation of the mixed anhydride with trifluoroacetic anhydride and Et_3N, followed by slow addition to triflic acid in dichloroethane (DCE) at reflux temperature (83 °C).(*29*) This addition protocol minimized intermolecular side reactions (running the reaction under high dilution conditions achieved the same result), leading to a 62-68% isolated yield of **9** following hydrolysis of the ethyl ester. The yield for this sequence did not suffer upon scale up (to ca. 3 kg of **9**), unlike the previous sequence.

Conclusions

The development of an efficient synthesis of **1** required investigation of a wide variety of synthetic issues. We have demonstrated that the preparation of clinical bulk supplies under cGMP conditions places a different and in general more stringent set of requirements on our skills as synthetic chemists. For example, while chromatographic purification is frequently acceptable in the synthesis of milligrams or even grams of material, this purification technique is impractical on multi-kilogram scale. Likewise, the use of toxic intermediates such as organomercurials can be dealt with on laboratory scale, but pose much more serious challenges on kilogram scale when preparing material slated for use in a clinical setting. These observations suggest that there remain many challenging areas in the arena of organic synthesis where the development of selective, low-cost, and environmentally-friendly methods will be of tremendous benefit to chemists in both academic and industrial fields.

References and Notes

1. Author to whom correspondence should be addressed: Pfizer Global Research & Development, P.O. Box 8156-101, Groton, CT, 06335, USA. Phone: (860) 441-6334, FAX: (860) 441-5445, e-mail: john_a_ragan@groton.pfizer.com.
2. Current address: Chemical Process Development, Merck & Co., Rahway, NJ.
3. (a) Davies, M. F In *The Pharmacology of the Gamma-Aminobutyric Acid System*; Baskys, A. and Remington, G., Eds.; Brain Mechanisms and Psychotropic Drugs; CRC: Boca Raton, FL, 1996; pp 101-116. (b) Krogsgaard-Larsen, P.; Froelund, B.; Joergensen, F. S.; Schousboe, A. *J. Med. Chem.* **1994**, *37*, 2489-2505.
4. (a) Eistert, B.; Haupter, F.; Schank, K. *Liebigs Ann. Chem.* **1963**, *665*, 55-67. (b) Maclean, I.; Sneeden, R. P. A. *Tetrahedron* **1965**, *21*, 31-34. (c) Hutmacher, H.-M.; Kruger, H.; Musso, H. *Chem. Ber.* **1977**, *110*, 3118-3125. (d) Suzuki, M.; Watanabe, A.; Noyori, R. *J. Am. Chem. Soc.* **1980**, *102*, 2095-2096. (e) Nishiguchi, I.; Hirashima, T. *Chem. Lett.* **1981**, 551-554. (f) Bushan, V.; Chandrasekaran, S. *Synth. Commun.* **1984**, *14*, 339-345. (g) Vankar, Y. D.; Chaudhuri, N. C.; Rao, C. T. *Tetrahedron Lett.* **1987**, *28*, 551-554.
5. Kneen, G.; Maddocks, P. J. *Synth.Commun.***1986**, 16; 13; 1986; 1635-1640.
6. Nagarajan, Kuppuswamy; Talwalker, Purnachard K.; Goud, A. Nagana; Shah, Rashmi K.; Shenoy, Sharada J.; Desai, Narasimha D. *Indian J.Chem.Sect.B* **1988**, 27; 1-12; 1113-1123.
7. (a) Norris, R. K.; Sternhell, S. *Aust. J. Chem.* **1972**, *25*, 2621. (b) Hodgson, H. H.; Nicholson, D. E. *J. Chem. Soc.* **1928**, 1879.
8. (a) Sawyer, J. S.; Schmittling, E. A.; Palkowitz, J. A.; Smith, W. J. *J. Org. Chem.* **1998**, *63*, 6338. (b) Suffert, J.; Raeppel, S.; Raeppel, F.; Didier, B. *Syn. Lett.* **2000**, 874. (c) Politanskaya, L.; Malykhin, E.; Shteingarts, V. *Eur. J. Org. Chem.* **2001**, *2*, 405. (d) Marriott, J. H.; Barber, A. M. M.; Hardcastle, I. R.; Rowlands, M. G.; Grimshaw, R. M.; Neidle, S.; Jarman, M. *J. Chem. Soc., Perkin Trans. 1* **2000**, *24*, 4265.
9. (a) Maryanoff, B. E.; McComsey, D. F.; Ho, W.; Shank, R. P.; Dubinsky, B. *Bioorg. Med. Chem. Lett.* **1996**, *6*, 333. (b) Beach, M. J.; Hope, R.;

Klaubert, D. H.; Russell, R. K. *Synth. Commun.* **1995**, *25*, 2165. (c) Davey, D. D.; Erhardt, P. W.; Cantor, E. H.; Greenberg, S. S.; Ingebretsen, W. R.; Wiggins, J. *J. Med. Chem.* **1991**, *34*, 2671. (d) Alo, B. I.; Avent, A. G.; Hanson, J. R.; Ode, A. E. *J. Chem. Soc., Perkin Trans. 1* **1988**, 1997.

10. [10] (a) Naito, Y.; Goto, T.; Akahoshi, F.; Ono, S.; Yoshitomi, H. *Chem. Pharm. Bull.* **1991**, *39*, 2323. (b) Quallich, G. J.; Morrisey, P. M. *Synthesis* **1993**, 51.
11. (a) Chiu, C. F.-F. *Review of Alkenyl and Aryl C-O Bond Forming Reactions*; Chiu, C. K.-F. Ed.; Pergamon: New York, **1995**; Vol. 2; pp 683-685. (b) Keegstra, J. A.; Peters, T. H. A.; Brandsma, L. *Tetrahedron* **1992**, *48*, 3633. (c) Lindley, J. *Tetrahedron* **1984**, *40*, 1433.
12. Bacon, R. G. R.; Rennison, S. C. *J. Chem. Soc. (C)* **1969**, 312.
13. Greene, T. W.; Wuts, P. G. M. *Protective Groups in Organic Synthesis*; Wiley: New York, NY, 1991; pp 309-405.
14. (a) Ragan, J. A.; Makowski, T. W.; Castaldi, M. J.; Hill, P. D. *Synthesis* **1998**, 1599. (b) Ragan, J. A.; Jones, B. P.; Castaldi, M. J.; Hill, P. D.; Makowski, T. W. *Organic Syntheses* **2001**, *78*, 63.
15. 2000-2001 Aldrich Handbook of Fine Chemicals.
16. (a) Corey, E. J. *J. Am. Chem. Soc.* **1953**, *75*, 2301. (b) Murabayashi, A.; Makisumi, Y. *Heterocycles* **1990**, *31*, 537.
17. Raphalen, A.; Sturtz, G. *Bull. Soc. Chim. Fr.* **1971**, 2962.
18. Tsuji, J.; Nagashima, H.; Hori, K. *Chem. Lett.* **1980**, 257-260.
19. Ragan, J. A.; Makowski, T. W.; am Ende, D. J.; Clifford, P. J.; Young, G. R.; Conrad, A. K.; Eisenbeis, S. A. *Org. Process Res. Dev.* **1998**, *2*, 379-381.
20. Krepski, L. R.; Hassner, A. *J. Org. Chem.* **1978**, *43*, 3173-3179.
21. Pak, C. S.; Kim, S. K.; Lee, H. K. *Tetrahedron Lett.* **1991**, *32*, 6011-6014.
22. (a) Pak, C. S.; Kim, S. K. *J. Org. Chem.* **1990**, 55, 1954-1957. (b) Brady, W. T.; Lloyd, R. M. *J. Org. Chem.* **1979**, *44*, 2560-2564. (c) Footnote 11 in ref 22(a) states that dechlorination of TMS enol ether **22** (Bu_3SnH) followed by desilylation (Bu_4NF) does indeed form cycloheptane-1,3-dione. No experimental details are provided.
23. Noyori, R.; Hayakawa, Y. *Org. React.* **1983**, *29*, 163-344.
24. (a) Honda, T.; Ishikawa, F.; Kanai, K.; Sato, S.; Kato, D.; Tominaga, H. *Heterocycles* **1996**, *42*, 109-112. (b) Molander, G. A.; Carey, J. S. *J. Org. Chem.* **1995**, *60*, 4845-4849. (c) Frimer, A. A.; Weiss, J.; Gottlieb, H. E.; Wold, J. L. *J. Org. Chem.* **1994**, *59*, 780-792.
25. Li, B.; Kasthurikrishnan, N.; Ragan, J. A.; Young, G. R. *Org. Process Res. Dev.* **2002**, *6*, 64-66.

26. House, H. O.; Wasson, R. L. *J. Am. Chem. Soc.* **1956**, *78*, 4394-4400.
27. Inoue, M.; Frontier, A. J.; Danishefsky, S. J. *Angew. Chem., Int. Ed.* **2000**, *39*, 761-764.
28. Although moderately expensive on research scale ($2-3/g from Alrich), 3-furoic acid was available to us as a raw material from an unrelated development candidate, and we felt that the price would likely decrease if larger orders were placed in the future.
29. Similar cyclization conditions were utilized by workers at Sakai Research Laboratories: Chujo, I.; Masuda, Y.; Fujino, K.; Kato, S.; Mohri, S.; Ogasa, T.; Kasai, M. *Abstracts of Papers*, 217th National Meeting of the American Chemical Society, Anaheim, CA.; American Chemical Society: Washington, DC, **1999**; ORG 77.

Chapter 4

Process Research and Initial Scale-Up of ABT-839: A Farnesyltransferase Inhibitor

Todd S. McDermott[1], Ramiya Premchandron[1,2], Anne E. Bailey[1], Lakshmi Bhagavatula[1], and Howard E. Morton[1]

[1]GPRD Process Research and Development, Department R450, Abbott Laboratories, 1401 Sheridan Road, North Chicago, IL 60064
[2]Current address: Process Research and Development, MediChem, 12305 South New Avenue, Lemont, IL 60439

The antitumor compound ABT-839 was synthesized in eight steps from commercially available starting materials. The synthesis was carried out with a single purification prior to final salt isolation. The key sequence is an efficient bromination and a novel bis-coupling strategy that transforms the biaryl core to the penultimate ester in a single operation with no isolation in 94% yield. After bisulfate salt formation, the product was isolated as a crystalline solid. The synthesis was carried out on kilogram scale and gave an overall yield of 43%.

Introduction

Inhibitors of farnesyltransferase (FT) have been of interest for their antitumor activity against a number of human tumor cell lines (*1-3*). The mode of action of these compounds in not entirely clear, but they have been shown to be effective at blocking or reversing tumor growth (*1*). **ABT-839** is a farnesyltransferase inhibitor under investigation for the treatment of various types of cancer (*2,3*).

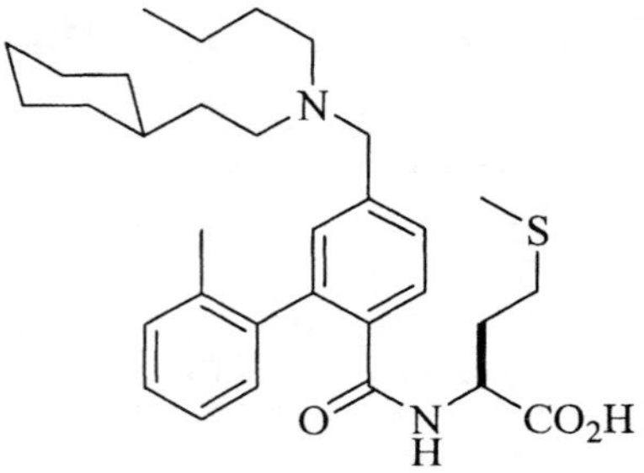

ABT-839

The original synthesis of **ABT-839** (*2*) efficiently provided material for early studies and provided the flexibility to vary the sidechains in order to study SAR. However, the route required at least three chromatographic separations, which prohibited synthesis of kilogram quantities of material. The penultimate intermediate was purified by column chromatography prior to the final ester hydrolysis and the final product was isolated as an amorphous foam. We were seeking an efficient process that addressed the concerns of purification of intermediates, allowed for isolation of a crystalline final product and could be used to synthesize kilogram quantities of material.

Process Synthesis of ABT-839

Retrosynthetic analysis of **ABT-839** (Figure 1) leads to the biaryl core, **1**, and the two sidechains; *N*-butyl-*N*-(cyclohexylethyl)amine (**2**) and methionine methyl ester (**3**). Acid **1** can be derived from diester **4**, which could come from Suzuki coupling of commercially available boronic acid **5** and dimethyl iodoterephthalate (**6**). This retrosynthesis features essentially the same disconnections used in the original synthesis, however, it was important that this route proceed through intermediate **1** since this was found to be a highly crystalline compound and its purification could be easily carried out thereby avoiding chromatographic purifications that might impede throughput.

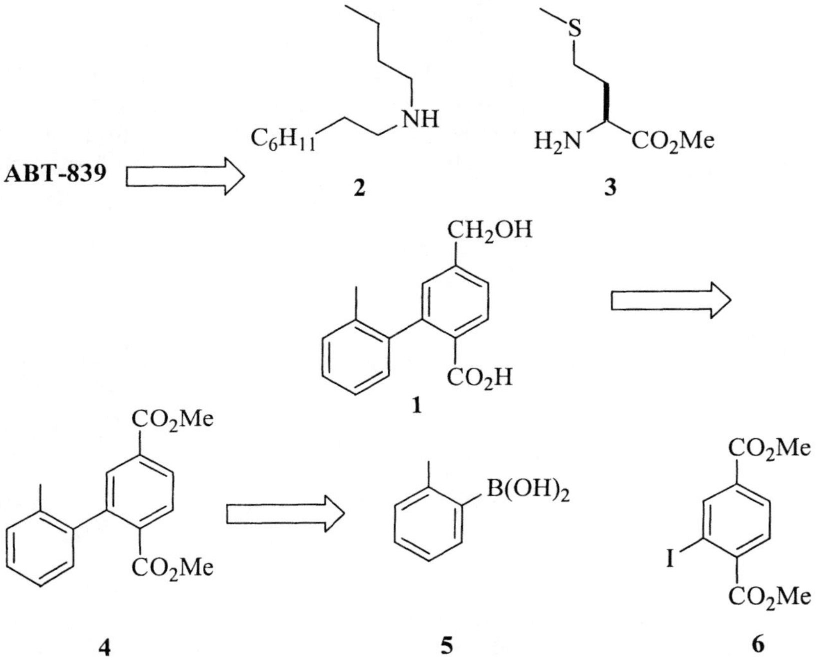

ABT-839

C_6H_{11} NH

2

H_2N CO_2Me S

3

CH_2OH

CO_2H

1

CO_2Me

CO_2Me

4

$B(OH)_2$

5

CO_2Me

I

CO_2Me

6

*Figure 1. Retrosynthetic analysis for **ABT-839**.*

Preparation of the Biaryl Core

A Suzuki cross-coupling reaction is employed to construct the core of **ABT-839**. The highly activated dimethyl iodoterephthalate (**6**) efficiently couples to *o*-toluyl boronic acid (**5**) under a variety of conditions (*4*). Catalysts ranging from palladium on carbon (*5*) to $Pd(PPh_3)_4$ effectively produced the desired product. In practice, the reaction was carried out in the presence of 1 mol% $Pd(OAc)_2$ and 4 mol% PPh_3 in the typical fashion with toluene as the organic phase and 2 M sodium bicarbonate as the aqueous base.

The major concerns with this reaction were dimerization of **6** to give tetra-ester **7** and removal of residual catalyst and unreacted boronic acid. Many attempts were made to find an efficient method to separate dimer **7** from diester **4**, but short of column chromatography, this separation was not possible. The source of this product was oxygen contamination of the solvents used for the reaction. Since removal of the byproduct was difficult, careful deoxygenation of the solvents was necessary to insure that the dimerization product was not formed. Prior to heating to 75 °C neither the Suzuki reaction nor the dimerization reaction proceed at all. This allowed all components of the reaction, including the solvent, to be premixed and sparged with nitrogen for 20-40 min prior to heating. This protocol completely suppresses the formation of **7**. Ester **4** can be purified by crystallization, but because the Suzuki reaction is typically very clean, purification at this early stage in the synthesis was unnecessary.

5 + **6** $\xrightarrow{Pd(OAc)_2,\ PPh_3}$ toluene, 2 M Na_2CO_3 (aq), 75 °C (99%) → **4** + **7**

CO_2Me CO_2Me MeO_2C CO_2Me MeO_2C CO_2Me

A work-up was devised to remove the catalyst and most remaining starting material without the need for isolation. Upon cooling to room temperature, the layers were separated and the organic layer was filtered through a plug of silica gel, which was then thoroughly washed with MTBE. Upon standing, a fine white precipitate (presumably a boron-containing byproduct) formed and was removed by filtration through Celite® with MTBE. Subsequent work has shown that the second filtration can be avoided by washing the crude toluene layer from the reaction with 10% aqueous triethanolamine prior to the silica gel filtration. The modified work-up up has been demonstrated on 100 g scale. The Suzuki reaction was carried out in 2 runs of approximately 2.5 kg each to produce a total of 4.4 kg of ester **4** (99% yield).

Differentiation of the Ester Groups

The original synthesis of **ABT-839** utilized a selective hydrolysis of the less-hindered ester followed by borane reduction of the resulting acid. An alternative approach would be to exploit the different steric environments of the two esters to reduce the accessible ester selectively to the corresponding alcohol. Of the common reducing agents tried, some showed little or no conversion (DIBAL-H in THF, BH_3•SMe_2, and $NaBH_4$), and others were non-selective or gave over-reduction (DIBAL-H in toluene or CH_2Cl_2, $LiBH_4$ in THF, and potassium tri-*sec*-butylborohydride (K-Selectride®)). Both lithium tri-*sec*-butylborohydride (L-Selectride®) and lithium triethylborohydride (Super-Hydride®) gave clean, fast reduction of **4** to alcohol **8** with little over-reduction (to give **9**) or reduction of the more hindered ester (to give **10a**). Removal of the trialkylborane byproducts of these reductions was easily addressed on gram scale, but presented some difficulty upon scaling. A common method to deal with the alkyl boranes is to oxidize the borane to the corresponding boronate ester by heating with alkaline hydrogen peroxide in MeOH (*6.7*). This procedure resulted in some oxidation of the benzylic alcohol **8** to the corresponding aldehyde. Even small amounts of this aldehyde proved difficult to remove during the subsequent purification. The original hydrolysis/reduction sequence was therefore reinvestigated.

4 $\xrightarrow{H^-}$ **8** + **9** + **10a, R = Me**

The diester **4** was treated with aqueous lithium hydroxide solution at 0 °C in THF/MeOH to give predominately the mono-acid product, along with small amounts of the corresponding di-acid and regioisomeric acid/ester. The crude mixture was treated with BH_3•SMe_2 in MTBE at 0 °C, followed by MeOH quench and aqueous NaOH wash to give ester **8** along with diol **9** (2.5 area%) and regioisomer **10a** (1-2 area%). Ester **8** is an oil and all attempts to induce crystal formation failed. Compound **8** could be separated from diol **9** by column chromatography, however, regioisomer **10a** co-elutes with **8** by TLC and HPLC and no efficient means of separation was possible at this point. The hydrolysis/reduction sequence was carried out to produce 2.5 kg of **8** in 69% yield for the two steps. The crude MeOH solution of **8**, which was about 95 area% pure by HPLC analysis, was taken directly into the next step.

The methanol solution from the hydrolysis/reduction sequence was treated with aqueous NaOH at 75 °C to hydrolyze the remaining ester. The product mixture contained **1** along with diol **9** and **10b**. After distillation of the MeOH, the aqueous layer was extracted with *i*-PrOAc to remove diol **9**. Acidification and extraction of the product with EtOAc followed by a solvent switch to toluene resulted in crystallization of **1**.

4 —a; b (69%)→ **8** (CO_2Me, HO) + **9** + **10a**

Reagents: (a) aq. LiOH, THF/MeOH, 0 °C; (b) BH_3•SMe_2, MTBE; aq NaOH.

This is a key point in the synthesis because the regioisomer **10b** is effectively removed and **1** is obtained in greater than 99 wt% purity. The hydrolysis and crystallization sequence proceeds in 87% yield and is the only purification during the first four steps of the synthesis.

8 —Aq NaOH, MeOH, 75 °C (87%)→ **1** (CO_2H, HO) + **9** + **10b, R = H**

Synthesis of Amine Sidechain

The HCl salt of *N*-butyl-*N*-(cyclohexylethyl)amine (**2•HCl**) was synthesized from commercially available cyclohexylacetic acid (**11**). Acid chloride formation using thionyl chloride in toluene with catalytic DMF was followed by amine acylation under Schotten-Baumann conditions to give the amide (*8*). After workup, the resulting toluene solution was azeotropically dried and the amide was reduced with BH_3•SMe_2. The crude reaction mixture was quenched with HCl in MeOH and the resulting salt was crystallized from toluene/MTBE. The salt was isolated in 80% yield for the 3 steps.

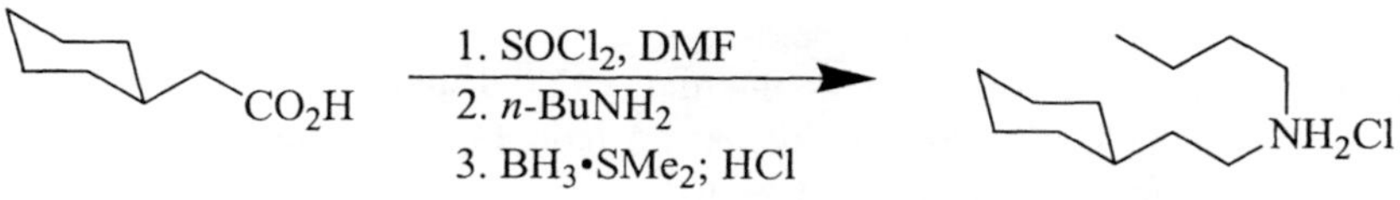

$$\text{11} \xrightarrow[\substack{3.\ BH_3\bullet SMe_2;\ HCl \\ (3\ steps,\ 80\%)}]{\substack{1.\ SOCl_2,\ DMF \\ 2.\ n\text{-}BuNH_2}} \text{2-HCl}$$

Functionalization of the Biaryl Core

The benzylic alcohol functionality of **1** precludes activation of the carboxylic acid for coupling to methionine. Rather than employing a protecting group to allow flexibility in the coupling strategy, the alcohol was transformed directly to benzylic bromide **12**. A suspension of **1** in toluene was treated with 48% aqueous HBr solution and the mixture was thoroughly deoxygenated by sparging with nitrogen. The sparging is necessary in order to avoid oxidation of **1** to the corresponding aldehyde, which is very difficult to remove during subsequent steps. As the mixture was heated, the solid slowly dissolved to give two completely clear layers at the completion of the reaction. The work-up simply involved separating the layers and washing the organic layer with water to remove residual hydrobromic acid, followed by azeotropic drying of the solution by toluene distillation. The reaction was carried out on 2 kg scale to give quantitative yield of product **12** in purity greater than 99 area% by HPLC analysis. The toluene solution was used directly in the next reaction.

$$\text{1} \xrightarrow[\substack{Toluene,\ 75\ °C \\ (100\%)}]{48\%\ Aq\ HBr} \text{12}$$

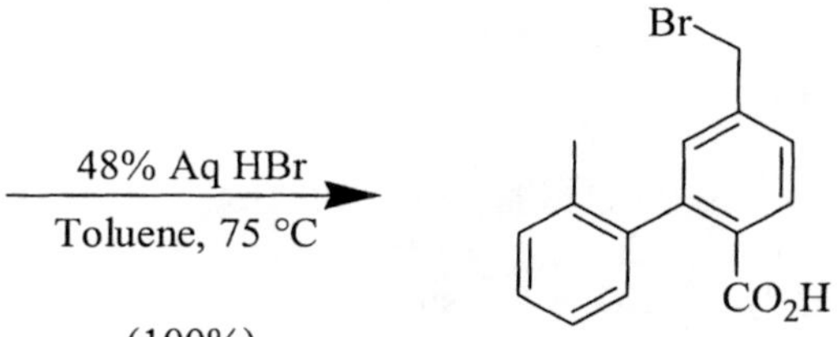

One of the aspects of the original synthesis that we were trying to avoid was the use of a carbodiimide-coupling reagent to couple the methionine methyl ester to the biaryl core. The reaction was straightforward to run, but the product was not crystalline and had to be purified by chromatography to remove impurities formed during the reaction. An alternative route that we investigated early on was the use of an acid chloride, however, the timing of the reaction was tricky. When the amine sidechain was installed first, an acid chloride coupling was problematic. On the other hand, the methionine methyl ester could be coupled to the acid chloride **13** without incident, but attempts to isolate **14** resulted in polymerization, probably due to intermolecular displacement of the benzylic bromide by the sulfide of methionine. Amide **14** was found to be stable in solution for an extended time. This led to the realization that the methionine coupling and the secondary amine S_N2 reaction could be carried out successively in one pot.

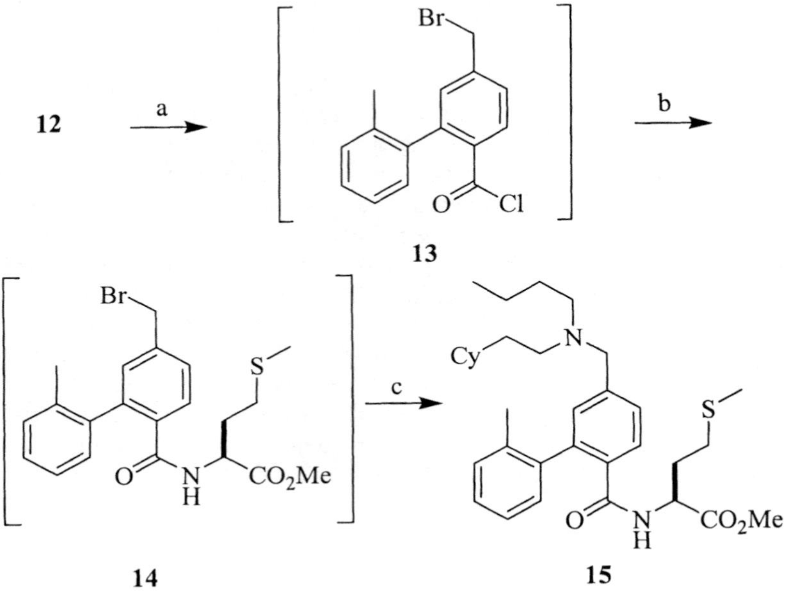

Reagents: (a) oxalyl chloride, cat. DMF, toluene, 1h; (b) methionine methyl ester·HCl; Et_3N, toluene, 0 °C, 10 min; (c) **2·HCl**, K_2CO_3, toluene/CH_3CN/H_2O, 8 h (3 steps, 94% yield).

Treatment of the toluene solution of bromide **12** with oxalyl chloride and catalytic DMF at room temperature led to the formation of acid chloride **13**. This acid chloride was stable in toluene at room temperature. The toluene solution was concentrated to half volume under reduced pressure in order to remove excess oxalyl chloride. This was necessary to avoid formation of oxamide **16**, a result of the addition of methionine methyl ester to oxalyl chloride. Because **16** proved difficult to remove, it was necessary to verify that the toluene concentration had completely removed the excess oxalyl chloride. This was accomplished by quenching an aliquot of the reaction mixture into a dilute solution of phenethylamine and assaying for the amounts of phenethyloxamide **17** and phenethylamide **18** versus standards. The toluene solution of **13** was cooled to 0 °C and treated with solid methionine methyl ester hydrochloride, which was then free-based by the dropwise addition of Et_3N. The coupling reaction, to give **14**, was complete as soon as all the Et_3N was added. Because methionine methyl ester readily adds to the benzylic bromide under the conditions of the S_N2 reaction, it was necessary to remove any excess methionine methyl ester prior to the next reaction. Therefore, the toluene solution was washed with dilute aqueous H_3PO_4 solution and returned to the reaction flask prior to treatment with solid **2•HCl**, aqueous K_2CO_3 and CH_3CN. The resulting mixture was sparged with nitrogen to avoid oxidation of the methionine sulfide to the sulfoxide, and the biphasic mixture was stirred for 8-10 h at rt.

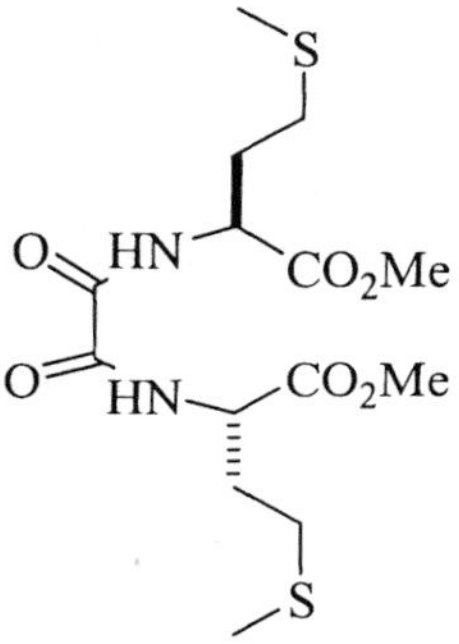

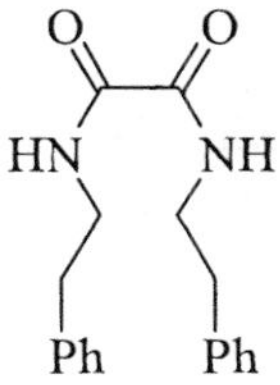

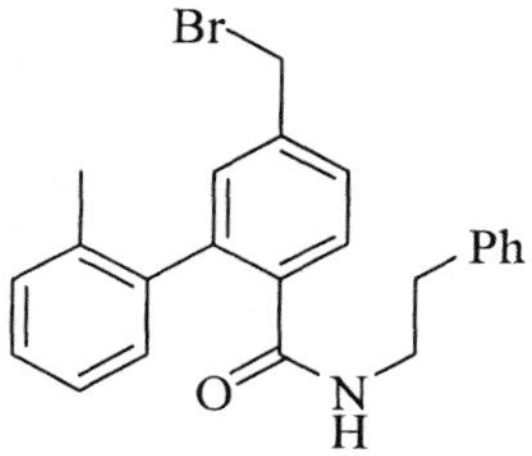

During the workup it is important to remove excess **2** from the product since it was difficult to remove during the subsequent step and remained as a contaminant after salt formation and final isolation of **ABT-839**. Its removal was accomplished by separating the layers and distilling the organic layer to remove the CH_3CN, followed by washing the toluene solution with 1 M AcOH. The acid wash effectively removes **2•AcOH** without loss of product to the aqueous layer (less than 0.1% **1** was detected in the aqueous layer). The product was free-based by washing with 2 M K_2CO_3 followed by a water wash and azeotropic drying. The procedure was carried out in 2 runs to produce about 4 kg of **15** in 94% yield. The material produced was 97 area% by HPLC analysis and chiral HPLC showed the product to be greater than 99.8% of the desired (*S*) enantiomer. The toluene solution was taken directly into the next reaction.

Final Hydrolysis and Salt Formation

Compound **15** was converted to the free base of the final drug by hydrolysis of the methyl ester. Careful control of the solvent ratio and temperature allowed the hydrolysis to be carried out with little or no epimerization of the stereocenter. The final challenge in the synthesis of **ABT-839** was identifying a pharmaceutically acceptable salt that allowed for purification of the final product. **ABT-839** was treated with a variety of acids and bases. Initially the potassium carboxylate looked promising in that it is a crystalline compound and its isolation resulted in substantial purification of the product, however, that salt is extremely hygroscopic. All other carboxylate salts investigated were non-crystalline. **ABT-839** was treated with a variety of acids. Of those investigated only the bisulfate salt of **ABT-839** crystallized. MTBE was the optimal solvent for the salt formation and crystallization, but great care had to be taken to deoxygenate the solvent prior to addition of the sulfuric acid in order to avoid oxidation of the sulfide to the sulfoxide. The hydrolysis/salt formation sequence was proceeded in 78% yield for the two steps and was utilized to produce 2 kg of **ABT-839•H_2SO_4**. Chiral HPLC analysis showed the product to be 99.4% of the desired (*S*) enantiomer.

15 → 1) aq LiOH, 0 °C; 2) H_2SO_4, 2-butanone (78%) → **ABT-839•H_2SO_4**

C_6H_{11} N •H_2SO_4 S O N H CO_2H

Conclusion

The synthesis of **ABT-839•H_2SO_4** was completed in eight chemical operations (10 steps total) in an overall yield of 43%. This synthesis was used to produce 4 kg of **ABT-839** free amine and 2 kg of **ABT-839•H_2SO_4**. The general approach was to utilize disconnections similar to those used in the original synthesis with an emphasis on streamlined work-up procedures and a single purification prior to final isolation. The steps for appending the two sidechains to the biaryl core were run in a single reactor and produced the penultimate product in 94% yield. Finally, the bisulfate salt provided an opportunity to purify the final product and had acceptable physical properties.

References

1. Prendergast, G. C. *Current Opinion in Cell Biology* **2000**, *12*, 166.
2. O'Connor, S. J.; Barr, K. J.; Wang, L.; Sorensen, B. K.; Tasker, A. S.; Sham, Ng, S.; Cohen, J.; Devine, E.; Cherian, S.; Saeed, B.; Zhang, H.; Lee, J. Y.; Warner, R.; Tahir, S.; Kovar, P.; Ewing, P.; Alder, J.; Mitten, M.; Leal, J.; Marsh, K.; Bauch, J.; Hoffman, D. J.; Sebti, S. M.; Rosenberg, S. H. . *J. Med. Chem.* **1999**, *42*, 3701.
3. Henry, K. J. Jr.; Wasicak, J.; Tasker, A. S.; Cohen, J.; Ewing, P.; Mitten, M.; Larsen, J. J.; Kalvin, D. M.; Swenson, R.; Ng, S-C.; Saeed, B.; Cherian, S.; Sham, H.; Rosenberg, S. H. *J. Med. Chem.* **1999**, *42*, 4844.
4. McDermott, T. S.; Bailey, A. E.; Premchandran, R.; Bhagavatula, L. Synthetic Preparations of Farnesyltransferase Inhibitors. U.S. Patent 6,248,919 B1, **2001**.
5. Buchecker, R.; Marck, G.; Billiger, A. *Tetrahedron Lett.* **1994**, *35*, 3277.
6. Brown, H. C. *Hydroboration;* W. A. Benjamin: Yew York, NY, **1962**; pp 69-72.
7. Brown, H. C. *Boranes in Organic chemistry*; Cornell University Press: Ithaca, .Y, **1972**; pp 321-325.
8. March, J. *Advanced Organic Chemistry*, 3rd ed.; John Wiley & Sons: New York, **1985**; p 370.

Chapter 5

Synthetic Approaches to the Retinoids

Margaret M. Faul

Lilly Research Laboratories, A Division of Eli Lilly and Company, Chemical Process Research and Development Division, Indianapolis, IN 46285–4813

Methods for the synthesis of retinoids (Acitretin, Etretinate, Isotretinoin, Tretinoin and Alitretinoin) derived from the tetraenoic acid platform via Wittig-Horner-Emmons reactions and photochemistry are described. Chemistry employed for synthesis of the more novel aryl/heteroaryl receptor selective retinoids (Bexarotene, Differin® and Zorac®) platforms is also reported.

Introduction

Retinoids, natural and synthetic analogs of retinol (vitamin A), play an important role in cell differentiation and vertebrate development, which appears to account for their therapeutic or preventative effects in acne, psoriasis, photoaging, cancer and other diseases.(*1*) Their activity is believed to be mediated through two classes of retinoid receptors:(*2,3*) the retinoic acid receptors (RAR) and the retinoid X receptors (RXR).(*4*) The RARs have a high affinity for both all-*trans*-retinoic acid and its isomer 9-*cis*-retinoic acid, the latter is also the natural ligand of the RXRs (Figure 1). Other retinoids having the 9,10-*cis* olefin geometry have been found to selectively activate the retinoid X receptor. The RXRs serve as ligand-dependent transcription factors and as heterodimerization partners for other members of the nuclear receptor superfamily including the RARs, the thyroid receptor (TR), the peroxisome proliferator-activated receptors (PPARs) and the vitamin D receptor (VDR). Due to their range of biological activities, the retinoids have demonstrated a large potential for inducing unwanted side effects including mucocutaneous toxicity, hypertriglyceridemia and teratogenesis.(*5*) Although retinoids have shown promise in dermatological and oncological indications, these adverse effects have hampered or restricted their use, particularly as preventive agents for chronic administration. Therefore one of the primary goals of retinoid research has been to design novel retinoids that have more favorable therapeutic indices with reduced risk of adverse effects and teratogenesis, and these efforts have been the subject of multiple reviews.(*6-8*) Recently the discovery of six nuclear retinoid receptors (RAR and RXR α,β, and γ) that mediate the biological effects of retinoids, (*9,10*) has allowed efforts to focus on development of receptor selective retinoids, that have a narrow range of adverse effects but maintain the specific therapeutic effects. In fact this effort has resulted in identification of two receptor-selective retinoids Tazarotene (Zorac®) and Adapalene (Differin®), topical drugs for treatment of psoriasis and acne (Figure 1).(*5, 11*)

Figure 1

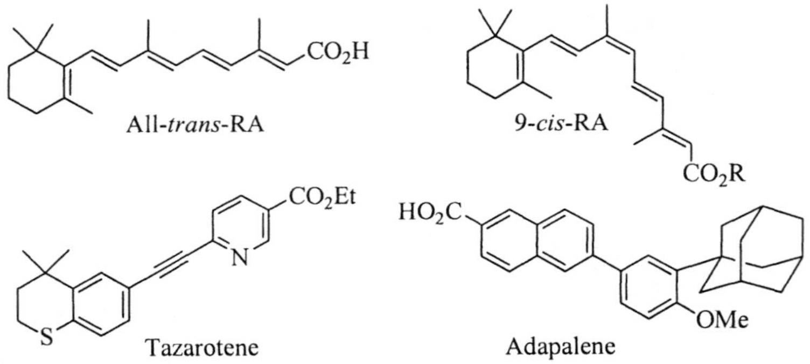

CO_2H

All-*trans*-RA

9-*cis*-RA

CO_2R

CO_2Et

N

S

Tazarotene

HO_2C

OMe

Adapalene

Retinoids currently on the market are outlined in Table 1. The goal of this article is to evaluate the processes employed for preparation of these compounds, since no review of this work has previously been compiled. The synthetic routes presented, identified by evaluation of publications and patents, are believed to be those performed on commercial scale. A comparison to the routes used pre-clinically is difficult due to the problem in obtaining this information in the literature. Organometallic approaches to the tetraenes although published, will not be documented due to limitations with the length of this article and issues associated with performing these reactions on commercial scale. The chemistry outlined should prove valuable in designing the next generation of retinoids.

Table 1: Commercial Retinoids and their Indications

Systemic Retinoids	*Company*	*Indication*
Acitretin (Soriatane®)	Roche	Severe psoriasis
Etretinate (Tegison®)	Roche	Severe recalcitrant psoriasis
Isotretinoin (Accutane®)	Roche	Severe recalcitrant cyctis psoriasis
Tretinoin(Vesanoid®)	Roche	Acute promyelocytic leukemia
Topical Retinoids	***Company***	***Indication***
Adapalene (Differin®)	Galderma	Acne vulgaris
Alitretinoin (Panrexin®)	Ligand	Kaposi's sarcoma lesions
Bexarotene (Targretin®)	Ligand	Acne vulgaris
Tretinoin (Renova®)	Ortho-McNeil	Acne vulgaris
Tazarotene (Zorac®)	Allergan	Acne vulgaris
Tretinoin (Retin A®)	Ortho-McNeil	Acne vulgaris
Tretinoin (Vesanoid®)	Pharmascience	Acne vulgaris

Synthetic Strategies

Alitretinoin (Panrexin®, 9-*cis*-retinoic acid or 9-*cis*-RA)

Alitretinoin is marketed by Ligand Pharmaceuticals Inc. for treatment of several cancers including renal cell carcinoma, non-Hodgkin's lymphoma, and acute promyelocytic leukemia.(*12*) Early attempts to prepare 9-*cis*-RA by olefination of β-ionone **1** were non-selective. Reformatsky and Wittig reactions gave mixtures of isomeric olefins including 9-*cis*-RA (or a synthetic intermediate), which was isolated from the mixture by selective crystallization.(*13-16*) Several approaches to improve the isomeric selectivity in formation of the 9*Z*-aldehyde **4** were reported.(*17-19*) Selective formation of the Z-trisubstituted double bond was achieved via 1,4-conjugate addition of dimethylcuprate to nitrile **2**, prepared in 76% yield from the commercially available and inexpensive β-ionone **1** (Scheme 1). This conjugate addition proceeded with high selectivity to give intermediate **3** as 98/2 mixture of the *cis*/*trans* isomers. DIBAL reduction of *cis*-

3 afforded aldehyde **4** in 95% yield, however, the reaction was accompanied by about 10% isomerization to the undesired 9*E*-aldehyde.(*17*)

Scheme 1

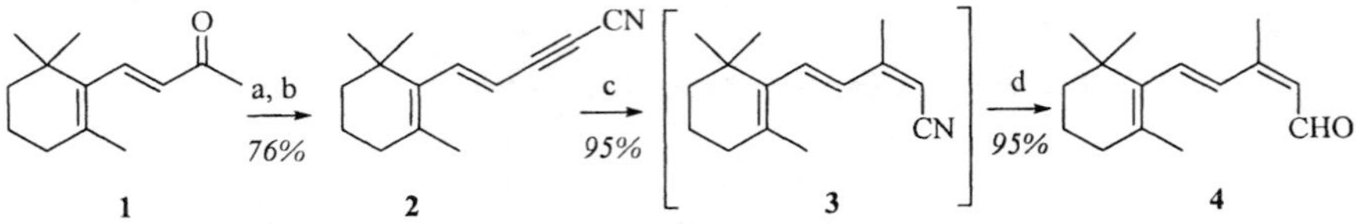

O; a, b; 76%; 1; CN; c; 95%; 2; CN; 3; d; 95%; CHO; 4

(a) (i) LDA, THF, -78 °C; (ii) $ClP(O)(OEt)_2$; (iii) LDA; H_2O; (b) *n*-BuLi, THF; (ii) PhOCN, -78 °C; (c) MeLi, CuI, THF, -78 °C; (d) DIBAL, hexanes.

Alternatively aldehyde **4** has been prepared by Reformatsky reaction of β-cyclocitral **5** with ethyl 4-bromo-3-methyl-2-butenoate **6** (Scheme 2). DIBAL reduction of lactone **7** afforded the lactol, which was opened with HCl to afford **4** in 75% yield. The success of this reaction is dependent on the pKa of the acid, the reaction concentration and the temperature.(*20*)

Scheme 2

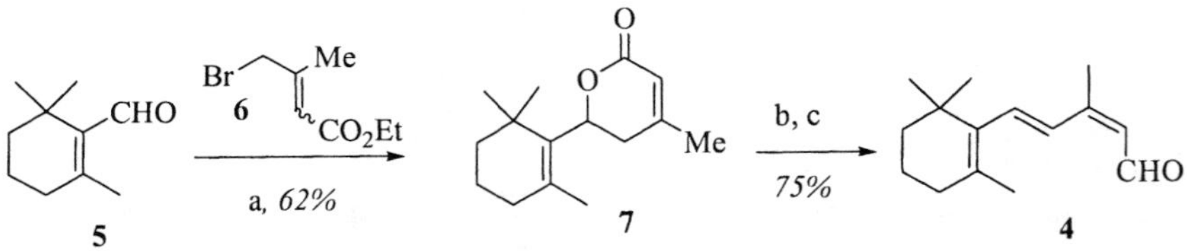

CHO; 5; Br; Me; 6; CO_2Et; a, 62%; O; O; Me; 7; b, c; 75%; CHO; 4

(a) Zn, THF; (b) DIBAL, THF; (c) HCl, $Cl_2(CH_2)_2$.

Wittig-Horner-Emmons reaction of **4** with diethyl 3-(ethoxycarbonyl)-2-methylprop-2-enyl phosphonate **8** afforded ester **9** in 89% yield as a 15:1 ratio of 13-*trans*/13-*cis* isomers. Saponification of **9** yielded 9-*cis*-RA in 75% yield (Scheme 3). This process via **1** completed a 6 step synthesis of 9-*cis*-RA in 50% yield, while the Reformatsky reaction sequence proceeded in 5 steps starting with **5** to give 30% overall yield. Both processes have been reported on kilogram scale.

Scheme 3

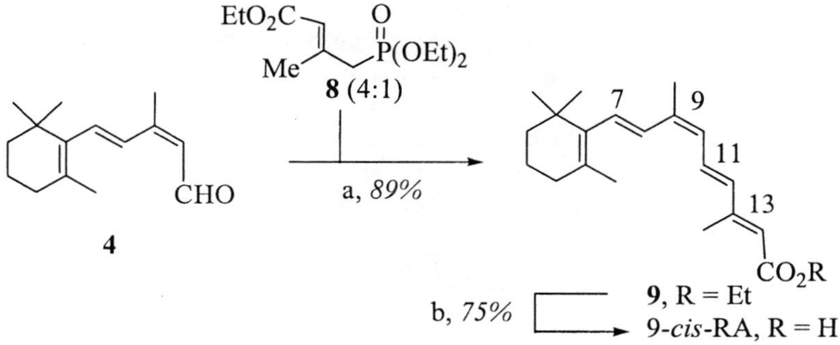

EtO_2C / Me / O / $P(OEt)_2$
8 (4:1)
CHO
4
a, *89%*
7 9 11 13
CO_2R
b, *75%*
9, R = Et
9-*cis*-RA, R = H

(a) *n*-BuLi, DMPU, THF, -40 °C; (b) KOH, EtOH, 70 °C.

Photochemical isomerization of all-*trans*-RA in acetonitrile using a tungsten filament lamp afforded material that contained 14% 9-*cis*-RA at equilibrium.(*21,22*) Recrystallization from EtOH provided 9-*cis*-RA in 5% yield (Scheme 4). Recycling of the mother liquor through photolytic re-equilibrium generated additional quantities of product (5% yield/cycle). Application of this chemistry to manufacturing scale was limited since the retinoic acids are not very soluble in organic solvents and the reaction needed to be performed at high dilution. Approaches to overcome these problems have been developed (*vide infra*).(*23-25*)

Scheme 4

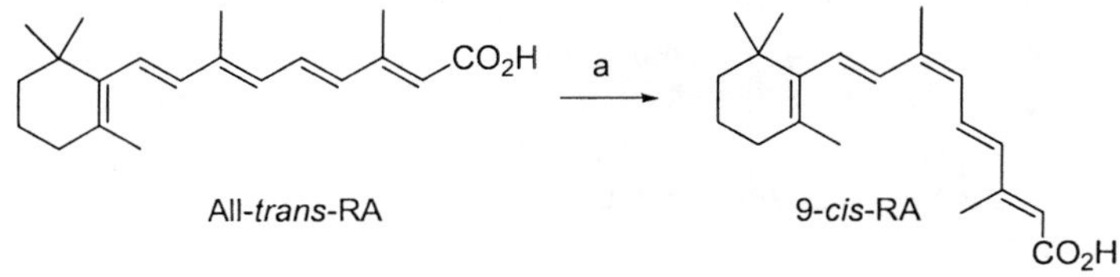

CO_2H
a
All-*trans*-RA
9-*cis*-RA
CO_2H

(a) hν, CH_3CN

Alternative synthetic approaches to 9-*cis*-RA have been published, although their application to commercial scale has not been demonstrated.(*18,26-29*)

Bexarotene (Targretin®)

Bexarotene is an RXR-selective retinoid marketed by Ligand Pharmaceuticals Inc., for treatment of refractory advanced-stage cutaneous T-cell lymphoma. Bexarotene is orally administered, safe and side effects are reversible.(*30*) The synthesis of Bexarotene starts with tetrahydronaphthalene **11**, prepared in 91% yield by Friedel-Crafts alkylation of toluene with 2,5-dichloro-2,5-dimethylhexane **10** (Scheme 5).(*31-33*) Friedel-Crafts acylation of **11** with chloromethylterphthalate **12** afforded ester **13** in 72% yield.

Scheme 5

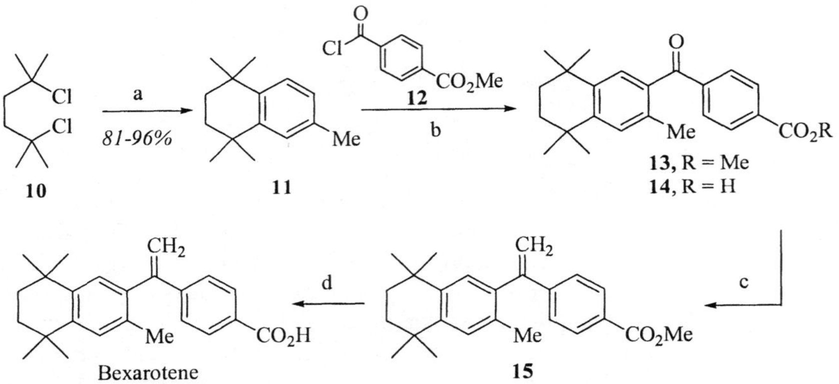

(a) Toluene (solvent), $AlCl_3$; (b) $AlCl_3$, CH_2Cl_2; (c) $MePPh_3Br$, $NaNH_2$, THF; (d) MeOH, KOH (aq), HCl.

Completion of the synthesis was achieved either (i) by olefination of **12** with methyltriphenylphosphonium bromide/$NaNH_2$ followed by saponification in 68% overall yield; or, (ii) by saponification of **13** to acid **14**, treatment with MeMgCl and dehydration with conc. HCl in 80% overall yield. The latter approach, which proved optimal, completed a 4 step synthesis of Bexarotene in 52% yield. Improvements to this process, involving catalytic Friedel-Crafts acylation chemistry have recently been reported.(*34*)

Tretinoin (Retin A,® Vesanoid®, Renova®, all-*trans*-RA)

The most general method to prepare all-*trans*-RA is by Wittig reaction of C_{15}-phosphonium salt **17**, prepared in two steps and 74% yield from commercially available β-ionone **1**, with methyl *E*-3-formylcrotonate (Scheme 6).(*23,35*) An alternative approach via the acetate has also been reported.(*36*)

Scheme 6

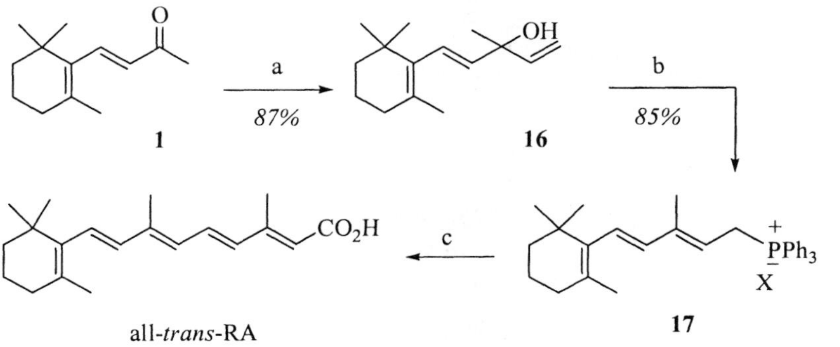

1 —a (87%)→ 16 —b (85%)→ 17 —c→ all-*trans*-RA

(a) Vinyl-MgBr, THF; (b) $PPh_3 \cdot HBr$, MeOH; (c) (i) *n*-BuLi, hexanes, methyl E-3-formylcrotonate; (ii) KOH, EtOH.

Tazarotene (Zorac®)

Tazarotene is an RAR selective retinoid, currently marketed by Allergan Inc., for the topical treatment of acne and psoriasis.(*37-39*) Two approaches to its synthesis have been reported (Scheme 7). Thiophenol **18** or **19** was treated with NaOH in the presence of 1-bromo-3-methyl-2-butene to afford the corresponding 1-arylthio-3-methyl-2-butene **20** or **21**, that upon cyclization with P_2O_5 and H_3PO_4 yielded benzothiopyran **22** or **23**. $SnCl_4$ catalyzed acylation of **22** with acetyl chloride afforded the 6-acetyl derivative, that upon dehydration with lithium diisopropylamide and diethyl chlorophosphate was converted into ethynyl benzothiopyran **24**. Alternatively **24** was prepared by Sonogishira coupling of bromobenzopyran **23** with TMS-acetylene, followed by removal of the TMS-protecting group. Completion of the synthesis was achieved by deprotonation of **24** with *n*-BuLi and reaction with ethyl 6-chloropyridine-3-carboxylate. Tazarotene was also prepared by palladium-catalyzed coupling of ethyl 6-chloropyridine-3-carboxylate with the organozincate derived from acetylene **24**. The process via acetylation is preferred on commercial scale since it eliminated the use of transition metals that contaminate the final product. No yield for either process has been reported in the literature.

Scheme 7

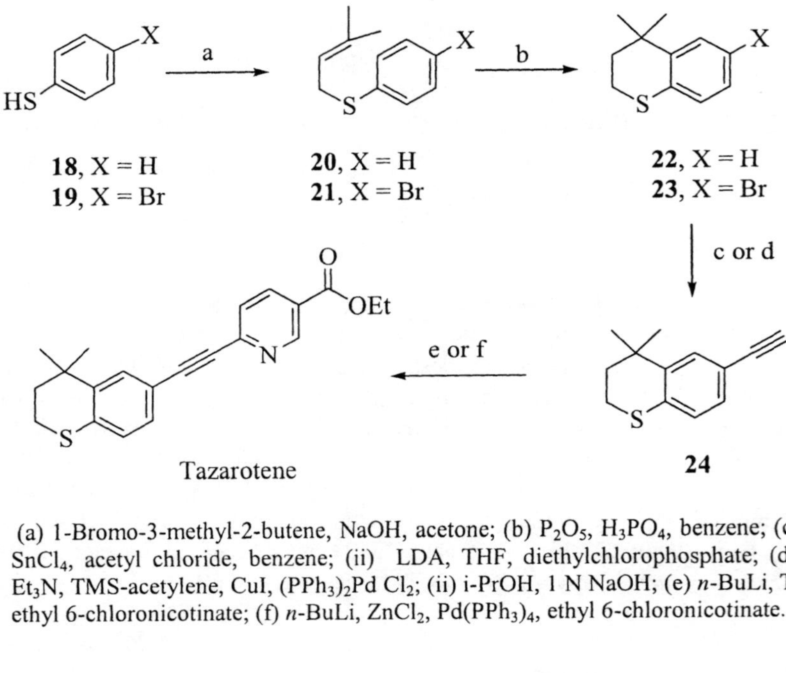

HS X a 20, X = H b X 22, X = H

18, X = H 21, X = Br 23, X = Br

19, X = Br

c or d

OEt

e or f

Tazarotene 24

(a) 1-Bromo-3-methyl-2-butene, NaOH, acetone; (b) P_2O_5, H_3PO_4, benzene; (c) (i) $SnCl_4$, acetyl chloride, benzene; (ii) LDA, THF, diethylchlorophosphate; (d) (i) Et_3N, TMS-acetylene, CuI, $(PPh_3)_2Pd\ Cl_2$; (ii) i-PrOH, 1 N NaOH; (e) *n*-BuLi, THF, ethyl 6-chloronicotinate; (f) *n*-BuLi, $ZnCl_2$, $Pd(PPh_3)_4$, ethyl 6-chloronicotinate.

Scheme 8

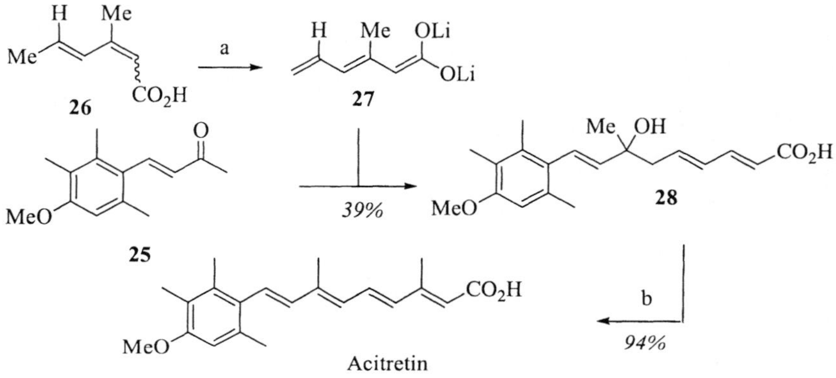

(a) LDA, THF; (b) H^+, I_2.

Acitretin (Soriatane®)

The major active metabolite of etretinate, is a second generation monoaromatic retinoid for use in the treatment of severe psoriasis and other dermatoses.(*40*) It is currently marketed by Roche Holding
AG. Two syntheses have been published. The first approach involved regio-selective addition of lithium trienediolate **27** (prepared, *in situ*, by treatment of (2*E*,4*E*)-3-methylhexa-2,4-dienoic acid **26** with LDA in THF) to the α,β-unsaturated ketone **25** (Scheme 8). Under equilibrium conditions the trienediolate added preferentially, through the ω-carbon, to ketone **25** in 1,2- or 1,4-fashion, yielding **28** in 39% yield. Acid catalyzed dehydration of **28** afforded acitretin in 94% yield. This procedure was also be applied to the synthesis of all-*trans*-RA.(*41*)

An alternative approach to the synthesis of acitretin has been reported from enone **25** (Scheme 9).(*42,43*) Reduction of **25** followed by addition of acetylene afforded alcohol **29** in 67% yield. Treatment of **29** with 2-methoxy-1-propene in the presence of *p*-TsOH, followed by reaction with NaOH afforded diene **30**. Treatment of **30** with *iso*-butyl chloroacetate yielded triene **32** via epoxide intermediate **31**. Treatment of **32** with PBr_3 and *N,N*-dimethylacetamide followed by ester saponification completed a synthesis of acitretin in 6 steps and 25% overall yield.

Scheme 9

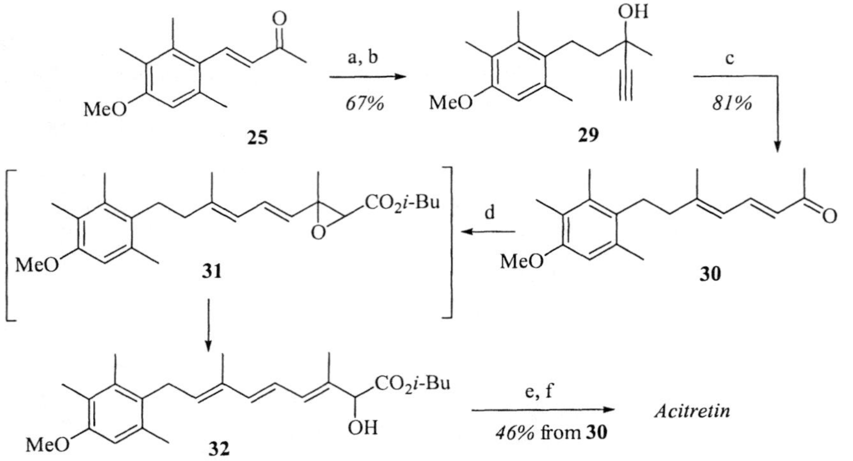

(a) H_2, Ra-Ni; (b) C_2H_2, NH_3, KOH, EtOH; (c) (i) 2-methoxy-1-propene, *p*-TsOH, 135 °C; (ii) NaOH, MeOH; (d) (i) *iso*-butyl chloroacetate, *iso*-BuOK; (ii) aq HCl; (e) (i) PBr_3, (ii) $MeCONMe_2$; (f) KOH.

Adapalene (Differin®)

Adapalene, is an RAR selective (β and γ) retinoid, marketed by CIRD Galderma, for the topical treatment of acne and psoriasis. It is a stable naphthoic acid derivative prepared by Friedel-Crafts alkylation of *p*-bromophenol **33** with 1-adamantanol to afford **34** in 86% yield (Scheme 10). Methylation of **34**, formation of the organozinc species and nickel-catalyzed cross condensation with methyl 6-bromo-2-naphthoate, afforded adapalene in 43% yield.(*44,45*)

Scheme 10

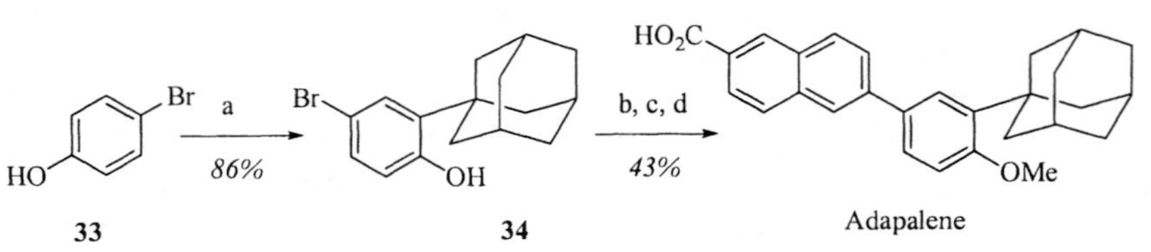

(a) 1-Adamantanol, H_2SO_4, CH_2Cl_2; (b) CH_3I, NaH, THF; (c) (i) Mg/THF; (ii) $ZnCl_2$; (iii) Methyl 6 bromo-2-naphthoate, $NiCl_2$/DPPE.

Isotretinoin (Accutane®, 13-*cis*-RA)

13-*cis*-RA, marketed by Roche Laboratories, is administered orally to treat dermatological diseases (severe, recalcitrant nodular acne).(*8*) The initial approach to the synthesis of 13-*cis*-RA, which contains an 11-*trans* double bond, involved a Wittig reaction of C_{15}-phosphonium salt **17** with C_5-butenolide **35** to afford a mixture of isomers (11-*trans*/13-*cis*-RA, 11-*cis*/13-*trans*-RA and 11-*trans*/13-*trans*-RA) from which the desired 13-*cis*-RA was purified by chromatography (Scheme 11).(*46*) Use of the C_{15}-phosphonate ester has also been explored but was uneconomical at commercial scale.(*47*)

Scheme 11

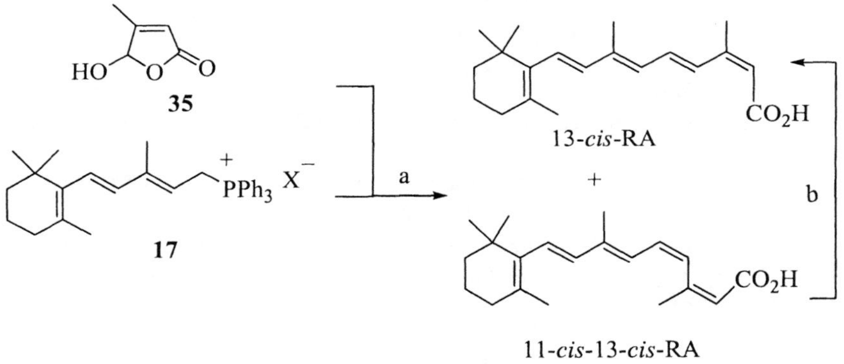

HO O O
35

$\overset{+}{P}Ph_3$ X^-
17

a

13-*cis*-RA
CO_2H
+
b
CO_2H
11-*cis*-13-*cis*-RA

(a) KOH; (b) $Pd(NO_3)_2$, PPh_3, TEA

Performing the Wittig reaction using a strong base, such as hydroxide or alkoxide, afforded >90% yield of product that contained 10-30% 11-*trans*/13-*cis*-RA and 70-90% of the 11-*cis*/13-*cis* isomer.(*48*) Upon treatment of this mixture with palladium or rhodium in an inert solvent, the 11-*cis* double bond selectively isomerized to the corresponding *trans* double bond without affecting the 13-*cis* double bond. However, application of this process to commercial scale was limited due to potential contamination of product with traces of the transition metal, potentially causing stability and toxcity issues with the polyene.

It has been reported that by performing the Wittig reaction in a polar aprotic solvent e.g. *N,N*-dimethylacetamide in the presence of triethylamine and magnesium chloride as Lewis acid, 13-*cis*-RA can be obtained in high selectivity (>98%). Unfortunately yields for this reaction were not reported.(*49*)

Photochemical isomerization of the mixture of isomers obtained from the Wittig reaction has been reported. Although this process avoids the use of transition metals, the retinoic acids are not very soluble in organic solvents and the reactions need to be performed at high dilution. However, using of the potassium or sodium salt of the retinoic acid and performing the reaction with a photoactivator, such as rose bengal, eliminated some of these issues and allowed the reaction to be carried out at shorter wavelengths (330 nM) than possible in organic solvents (353 nM).(*23-25*) In addition the alkaline salts of 11-*cis* and 13-*cis*-RA are more stable in aqueous solution than the corresponding acids allowing the reactions to be performed at higher concentrations. Thus photochemical isomerization of an isomeric mixture of retinoic acids composed of 23% 13-*cis*-RA, 64% 11-*cis*/13-*cis*-RA and 13% 11-*trans*/13-*trans*-RA, using the above protocol (KOH, H_2O, rose bengal) for 2 hr affords a mixture of 66% 13-cis-RA, 24% 11-*cis*/13-*cis*-RA and 10% 11-*trans*/13-*trans*-RA. After acidification and recrystallization from EtOAc a 53% yield of 13-*cis*-RA was obatined in >99% purity.(*23*)

A novel process for selective formation of the 13-*cis* double bond, by condensation of the dienolate of methyl 3-methyl-2-butenoate with β-ionylidene acetaldehyde **36** (9-*trans* content – 80%), has been reported (Scheme 12).(*50*) This reaction proceeded by formation of the intermediate lactone **37**, which was not isolated. Lactonization resulted in release of a methoxide ion that in turn opened the lactone to afford 13-*cis*-RA as the carboxylate salt. Acidic work-up, followed by recrystallization from MeOH produced 13-*cis*-RA in 32% yield, >99% purity (<0.1% all-*trans*-RA).

Scheme 12

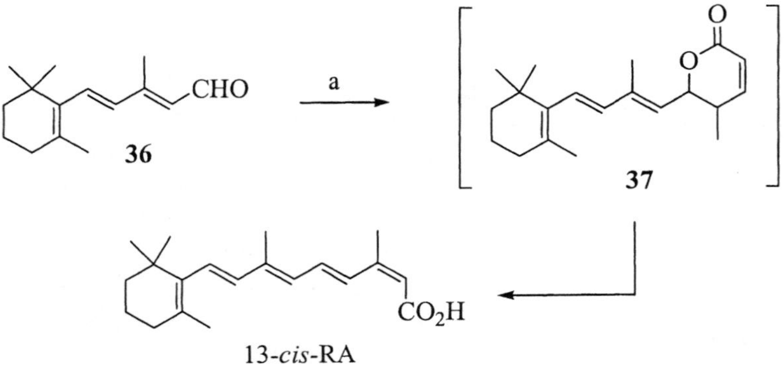

CHO a **36** **37** CO_2H 13-*cis*-RA

(a) (i) LDA, THF; methyl 3-methyl-2-butenoate; (ii) 10% sulfuric acid.

A number of other approaches to the synthesis of 13-*cis*-RA have been reported although they have not be demonstrated on commercial scale.(*51-54*)

Conclusion

The major approach for synthesis of retinoids of the tetraenoic acids involves the Wittig-Horner-Emmons reaction. The photochemical reaction has also proved viable and conditions to perform this reaction on commercial scale have been identified. Identification of receptor selective retinoids expanded the chemistry of the retinoids to aryl (Bexarotene, Adapalene) and heteroaryl (Tazarotene) platforms. These analogs are prepared rapidly and efficiently and this chemistry should facilitate both the SAR and commercial development of more novel and selective members of the retinoid class.

Acknowledgements

The author would like to thank Ms. Diane Desante for her contributions in compiling key references for the manuscript.

References

1. Gudas, L. J.; Sporn, M. B.; Roberts, A. B. In *Retinoids*; Sporn, M. B., Roberts, A. B., Goodman, D. S., Eds.; Raven Press: NewYork, 1994, p 443-520.
2. Gudas, L. J. *Cell Growth and Differentiation* **1992**, *3*, 655-672.
3. Leid, M.; Kastner, P.; Chambon, P. *Trends Biochem. Sci.* **1992**, *17*, 427-433.
4. Mangelsdorf, D. J.; Ong, E. S.; Dyck, J. A.; Evans, R. M. *Nature (London)* **1990**, *345*, 224-229.
5. Thacher, S. M.; Vasudevan, J.; Chandraratna, R. A. S. *Curr. Pharm. Des.* **2000**, *6*, 2-58.
6. Sun, S.-Y. *Expert Opin. Ther. Patents* **2002**, *12*, 529-542.
7. Curley, R. W.; Robarge, M. J. *Adv. Organ Biol.* **1997**, *3*, 1-34.
8. Nagpal, S.; Chandraratna, R. A. S. *Curr. Pharm. Des.* **2000**, *6*, 919-931.
9. Chambon, P. *FASEB J* **1996**, *10*, 940-954.
10. Piedrafita, F. J.; Pfahl, M. *Handbook of Experimental Pharmacology* **1999**, 153-184.

11. Shroot, B.; Gibson, D. F. C.; Lu, X.-P. *Handbook of Experimental Pharmacology* **1999**, *139*, 539-559.
12. Miller, W. H. J.; Jakubowski, A.; Tong, W. P.; Miller, V. A.; Rigas, J. R.; Benedetti, F.; Gill, G. M.; Truglia, J. A.; Ulm, E.; Shirley, M. *Blood* **1995**, *85*, 3021-3027.
13. Graham, W.; van Dorp, D. A.; Arens, J. F. *Rec. trav. chim.* **1949**, *68*, 609-612.
14. van Dorp, D. A.; Arens, J. F. *Rec. trav. chim.* **1946**, *65*, 338-345.
15. Arens, J. F.; van Dorp, D. A. *Nature* **1946.**, *157*, 190-191.
16. Robeson, C. D.; Cawley, J. D.; Weisler, L.; Stern, M. H.; Eddinger, C. C.; Chechak, A. J. *J. Am. Chem. Soc.* **1955**, *77*, 4111-4119.
17. Bennani, Y. L. *J. Org. Chem.* **1996**, *61*, 3542-3544.
18. Wada, A.; Hiraishi, S.; Takamura, N.; Date, T.; Aoe, K.; Ito, M. *J. Org. Chem.* **1997**, *62*, 4343-4348.
19. Wada, A.; Nomoto, Y.; Tano, K.; Yamashita, E.; Ito, M. *Chem Pharm Bull* **2000**, *48*, 1391-1394.
20. White, S. K.; Hwang, C. K.; Winn, D. T. **1994**, WO 9,424,082.
21. Coe, J. W.; O'Connell, T. *Bioorg. Med. Chem. Lett.* **1994**, *4*, 349-350.
22. Dawson, M. I.; Hobbs, P. D.; Rhee, S. W.; Morimoto, H.; Williams, P. G. *J Label Compd Radiopharm* **1993**, *33*, 633-637.
23. Magnone, G. A. **1999**, EP 959,069.
24. John, M.; Paust, J. **1994**, DE 4,313,089.
25. John, M.; Paust, J. **1995**, EP 659,739.
26. Wada, A.; Hiraishi, S.; Ito, M. *Chem Pharm Bull* **1994**, *42*, 757-759.
27. Pazos, Y.; de Lera, A. R. *Tetrahedron Lett.* **1999**, *40*, 8287-8290.
28. Iglesias, B.; Torrado, A.; de Lera, A. R.; Lopez, S. *J. Org. Chem.* **2000**, *65*, 2696-2705.
29. DeLuca, H. F.; Tadikonda, P. K. **1998**, US 5,808,120.
30. Duvic, M.; Hymes, K.; Heald, P.; Breneman, D.; Martin, A. G.; Myskowski, P.; Crowley, C.; Yocum, R. C. *J Clin Oncology* **2001**, *19*, 2456-2471.
31. Hurst, R. E. *Curr. Opin. Invest. Drugs* **2000**, *1*, 514-523.
32. Boehm, M. F.; Heyman, R. A.; Zhi, L. **1993**, WO 9,321,146.
33. Boehm, M. F.; Zhang, L.; Badea, B. A.; White, S. K.; Mais, D. E.; Berger, E.; Suto, C. M.; Goldman, M. E.; Heyman, R. A. *J. Med. Chem.* **1994**, *37*, 2930-2941.
34. Faul, M. M.; Ratz, A. M.; Sullivan, K. A.; Trankle, W. G.; Winneroski, L. L. *J. Org. Chem.* **2001**, *66*, 5772-5782.
35. Haugan, J. A. *Acta Chem Scandinavica* **1994**, *48*, 657-664.
36. Paust, J. *Pure Appl. Chem.* **1991**, *63*, 45-58.
37. Ngo, J.; Leeson, P. A.; Castaner, J. *Drugs of the Future* **1997**, *22*, 249-255.

38. Chandraratna, R., A. **1997**, U.S. 5,380,877.
39. Chandraratna, R. A. S. **1992**, US 5,089,509.
40. Pilkington, T.; Brogden, R. N. *Drugs* **1992**, *43*, 597-627.
41. Aurell, M. J.; Ceita, L.; Mestres, R.; Parra, M.; Tortajada, A. *Tetrahedron* **1995**, *51*, 3915-3928.
42. Zakharova, N. I.; Sokolova, N. N.; Gutnikova, N. P.; Muravev, V. V.; Filippova, T. M.; Rajgorodskaya, O. I.; Khristoforov, V. L.; Samokhvalov, G. I. **1993**, RU 2,001,903.
43. Samokhvalov, G. I.; Zakharova, N. I.; Sokolova, N. N.; Filippova, T. M.; Raigorodskaya, O. I.; Muravyev, V. V.; Tuguzova, A. M.; Khristoforov, V. L. *Khim-Farm. Zh.* **1994**, *28*, 43-47.
44. Charpentier, B.; Bernardon, J.-M.; Eustache, J.; Millois, C.; Martin, B.; Michel, S.; Shroot, B. *J. Med. Chem.* **1995**, *38*, 4993-5006.
45. Pfahl, M.; Lu, X.-P.; Rideout, D.; Zhang, H. **2001**, WO 156,563.
46. Pattenden, G.; Weedon, B. C. L. *J. Chem. Soc., C* **1968**, *16*, 1984-1987.
47. Babler, J. H. **1990**, US 4,916,250.
48. Lucci, R. **1985**, US 4,556,518.
49. Wang, X. C.; Bhatia, A. V.; Hossain, A.; Towne, T. B. **1999**, WO 9,948,866.
50. Salman, M.; Kaul, V. K.; Babu, J. S.; Kumar, N. **2001**, WO 109,089.
51. Thibonnet, J.; Abarbri, M.; Duchene, A.; Parrain, J.-L. *Synlett* **1999**, *1*, 141-143.
52. Wada, A.; Fukunaga, K.; Ito, M. *Synlett* **2001**, *6*, 800-802.
53. Valla, A.; Andriamialisoa, Z.; Giraud, M.; Prat, V.; Laurent, A.; Potier, P. *Tetrahedron Lett.* **1999**, *40*, 9235-9237.
54. Solladie, G.; Girardin, A.; Lang, G. *J. Org. Chem.* **1989**, *54*, 2620-2628.

Chapter 6

Design of Experiments in Pharmaceutical Process Research and Development

John E. Mills

Drug Evaluation – Chemical Development, Johnson & Johnson Pharmaceutical Research and Development, L.L.C., Welsh and McKean Roads, Spring House, PA 19440–0776

This chapter provides a brief overview of DOE. Discussion of the technique's advantages and limitations are provided. Nine examples of its use in pharmaceutical process research and development are included.

Background

For decades, statistics has been applied to problems related to quality control and process improvement in the manufacturing environment. The applicability of statistical methods in the chemical process research and development area has also been recognized for many years. Despite the advantages offered by a statistical treatment of the data, it has only been recently that the use of design of experiments (DOE) has started to become routine in early pharmaceutical process research and development.

Part of the driving force for acceptance of DOE has been the need to synthesize moderate quantities of new compounds that were initially produced using solid phase synthesis or other combinatorial techniques in a short time frame. Because of the low volume efficiencies and large excess of reagents typically associated with solid phase synthesis, direct scale up of those syntheses may not be practical. In this case, much of the information on intermediates previously available from medicinal chemists is no longer available because those intermediates are not isolated. The smaller amount of information available from discovery chemists who are first to prepare potential new drug entities, together with the short time frame available to produce materials needed for toxicology, physical pharmacy and formulation studies, have forced development chemists to examine their development process.

DOE offers the potential to maximize the information content of a series of experiments. The increased information content does, however, come at a cost. DOE experiments typically require advanced planning for a number of similar experiments. Even worse, in the minds of some chemists, is the fact that the designs require multiple repetitions of experiments. The coup de grâce for some in the arguments against the use of DOE is that it encourages simultaneous changes in multiple variables between experiments. This is contrary to conventional wisdom concerning well-controlled experimentation that states that only one variable should be changed at a time.

One Variable At a Time (OVAT) – The Classical Methodology for Process Development

Most organic chemists feel comfortable studying a new reaction by varying a single reaction parameter, such as temperature, at a time. This univariate design strategy is referred to as OVAT testing. Once the impact of variations in the first parameter are known, another reaction parameter, such as concentration of a reagent, may be varied, resulting in a multidimensional univariate study. In many instances, the results of one experiment are used to help design the next experiment. Under these constraints, the work is a sequential multidimensional univariate study. Such studies are slow and inefficient yet widely accepted.

Introductory laboratory manuals used in undergraduate chemistry help provide the basis for the bias toward OVAT experimentation. If experiments are included that are intended to explore the impact of a change in variables on the yield or product of a reaction, those experiments typically are designed to clearly show the effect of a single variable. For example, the reaction of semicarbazide with a mixture of cyclohexanone and 2-furaldehyde is known to produce different products depending upon whether the reaction is run under conditions that favor kinetic or thermodynamic control. Laboratory manuals that include this reaction typically instruct the student to run a series of reactions using the same ratio of reactants, in the same solvent, for short periods of time, and at different temperatures. The melting point of the isolated product can be used to determine which product is formed predominately. Although the experiment is a valuable educational tool because it provides experimental reinforcement of a number of chemical concepts, it also reinforces the concept that only by changing one variable at a time can one easily interpret the results from a series of experiments.

During the development of a new synthetic step using OVAT methodology, the chemist typically defines a set of conditions to be used as the starting point for the experiments. The first experiment is performed and a response (e.g. yield, purity, etc.) is determined. One of the parameters (Variable 1) in the first experiment is then changed and the response under the new condition is then determined. If the new response is not satisfactory, the two responses are compared and another experiment is designed in which Variable 1 is again modified with the intention of producing a new response that is more desirable than either of the previous responses. This process is repeated either until the desired response is obtained or until no further improvement in the response is observed. If no further improvement can be obtained by changing Variable 1, the chemist typically selects another reaction parameter (Variable 2) that can be systematically changed. The process of running a reaction under the new conditions, comparing the new response to the old responses, and systematically changing Variable 2 is continued until no further improvement in response due to a change in Variable 2 is possible or necessary. If the required response is still not met, another reaction parameter may be selected and the process is again repeated until 1) a set of conditions that do produce the desired response is found, 2) no additional controllable reaction parameters can be identified, or 3) time or resources for the study are expended, forcing acceptance of the best conditions found to date.

During the entire process, the chemist, perhaps without realizing it, has been using the concept of a response surface. In OVAT, one-dimensional responses are applied sequentially in order to reach the desired final response. If more than one variable is necessary to reach the desired response, then a portion of a multi-dimensional experimental domain has been examined. For simplification, further discussion of response surfaces in this chapter will be limited to two variables. The logic that applies to two-dimensional space can be extended to response surfaces in higher dimensions. Representation of dimensions higher than three is typically done mainly within computer memory.

In two dimensions, organic chemists' response surfaces are analogous to geologists' contour maps. Just as a geologist would examine a contour map to find locations of hills and valleys, chemists can use response surfaces to determine where (under what conditions) to maximize or minimize formation of a given product. Just as a geologist does not know the elevation of a location on a map in advance of a survey, the chemist does not know the yield of a specific product prior to running the reaction under those conditions. As discussed above, one of the challenges facing the process chemist is to locate an area within the experimental domain that has the desired response (normally high yield) with a minimal amount of effort.

In many experiments, each variable is not completely independent of all other variables. For example, a change in the temperature at which an experiment is performed may cause a change in the effect of the concentration of a reagent on the observed yield. In organic synthesis, such interactions tend to produce response surfaces such as that shown in Figure 1.

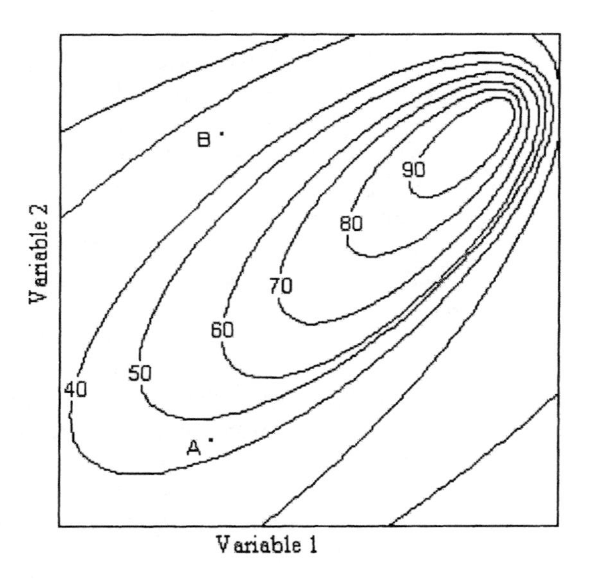

***Figure 1**. General representation of the effect on yield due to the interaction of two experimental variables.*

In order to see possible consequences of OVAT methodology, assume that the response surface in Figure 1 is the actual surface for a reaction. If the initial conditions are those given by Point A, then very little change in the response is found by either increasing (movement to the right) or decreasing (movement to the left) the value of Variable 1 by a small amount. An increase (movement upward) in the value of Variable 2 would lead to an improvement to about 55% yield, after which the yield would begin to drop. Upon completion of the series of experiments, one would conclude that the maximum yield that could be obtained was approximately 55%. If the first point to be selected was point B, the response would be seen to improve as Variable 1 was increased until it went through a maximum of about 90%. At this point, any changes in Variable 2 would tend to decrease the yield. In this case the conclusion that would be drawn is that the best yield that can be obtained is about 90%.

The first deduction that should be drawn from the example above is simple. In an OVAT series of experiments, the conclusions may be dependent upon the conditions used in the first experiment. Since the shape of the response surface is not known in advance of running the experiments, the results must be interpreted with caution. If the desired response is found during the study, the immediate goal may have been achieved. OVAT does not provide any meaningful information about the shape of the response surface except along the experimental path that was taken.

A second, though less obvious, deduction is that because OVAT experiments are essentially sequential they may require an extended period of time to reach a conclusion. New experimental conditions require the results of prior experiments to be known. One may be fortunate enough to find a set of reaction conditions that provide the desired response early in the process. Alternatively, dozens of experiments may be run over a long period of time without achieving the desired response, and without a good indication of whether or not it can ever be obtained.

Two typical goals in process development are to find the "best conditions" for a reaction and to make the reaction reproducible. Data produced using OVAT can make it difficult to generate a mathematical model describing the response of the system to changes in reaction parameters. Without such models, it may be difficult to predict how reproducible the reaction will be when scaled up.

DOE – An Effective Methodology to Improve the Information Content in a Series of Reactions

In advance of doing any laboratory work, it is possible to identify parameters (temperature, pressure, stirring rate, equivalents of reagents, etc.) that may affect the goal of a study (the desired response). It is also possible to select the mathematical model fitted to the generated data. For screening purposes a linear model is often selected in order to determine which parameters have the greatest impact upon the response. For production purposes it is useful to know if there are any interactions between parameters or higher order effects, therefore a quadratic model may be appropriate. Once the variables to be used in a study and the model to which the data is to be fit have been identified, a plan for generating the data can be constructed. The process of selecting the variables to be studied, selecting the model to which the data is to be fit, and building an efficient plan for data generation is referred to as design of experiments (DOE).

The effective use of DOE requires clearly defined goals before experiments are performed. Unlike OVAT, the minimum number of experiments required to construct and evaluate the model is known in advance. In addition, since the number of reactions and the desired setting for each reaction variable in each and every reaction is known at the start of a study, DOE provides the possibility of running multiple experiments simultaneously. The time saved by running parallel experiments can be significant. If, at the conclusion of a set of experiments, the desired response has not been obtained, DOE provides tools to evaluate the potential benefit of additional work and can be used to minimize the amount of additional experimental work that must be performed.

One of the best-known experimental designs is the two-level factorial design. In this approach, each variable is set to a high limit and a low limit defining the experimental domain. For convenience, the high limit is typically designated as +, and the low limit is typically designated as -. Each experiment is identified by a series of symbols designating the level of each experimental variable in the specific experiment. For two variables in a full two-level factorial experiment there is a minimum of four experiments that must be performed. One experiment is designated by (+,+) indicating that both variables are set to their upper limits. Another experiment is designated as (-,-) indicating that both variables are set to their low limits. The remaining two experiment are designated as (+,-) and as (-,+), indicating that the first factor is set to its high limit and the second factor is set to its low limit in one experiment and that the first factor is set to its low limit and the second factor is set to its high limit in the second experiment. The total number of experiments required in order to perform a full two-level factorial design experiment involving N factors is equal to 2^N. For experiments in which a small number of factors are under consideration, the total number of experiments appears reasonable. If, however, the number of factors is greater than five, the total number of experiments is greater than or equal to 64. This is more than most people would want to perform in an optimization using OVAT!

Fortunately, in most cases, a full two-level factorial design is not required. A full two-level factorial design with as few as four replicate experiments, makes it possible to construct a mathematical model of the system that includes terms to describe all primary effects and effects caused by the interactions of multiple variables. Experience has shown that, in chemistry, interactions of more than two variables are rare. Consequently, a model of that complexity is seldom required. In order to construct a less complex model with fewer parameters, less information is required. Fewer experiments need to be performed and fractional factorial designs may possibly be used. The main disadvantage of fractional factorial designs is that it may not be possible to unambiguously identify the source of an effect. The assignment of multiple parameters to the same mathematical term in a model is called aliasing. Care must be taken in designing fractional factorial experiments to minimize the probability of misinterpretation of a model due to aliasing. Commercial DOE software may provide the alias structure (i.e., explicitly name the mathematically equivalent parameters) in a given design. In order to gain additional information from a study while minimizing the number of additional experiments that must be performed, experiments can be run at the center of the experimental domain. Such experiments provide information on any curvature that should be included in the mathematical model of the response surface. As indicated above, a minimum of three replicate experiments should be included in the design in order to estimate the reproducibility of experiments within the experimental domain. Spreading the repetitions across as much of the experimental domain as possible can be advantageous.

The error observed in a given trial may be caused by one or all of the following factors:

1. The variation from the programmed value in the amount of any material added to a reaction.
2. The variation in temperature around its set point over the entire course of the reaction.
3. Variability in transfer volumes during sample preparation and analysis.

The shape of the response surface in close proximity to a given set of reaction conditions can produce wide variations in the response resulting from the above errors. If response contours are much more closely spaced in close proximity to one trial than to a second trial, then the observed error in repeating the first trial will be much greater than that for the second. By spreading repetitions over several different trials, one is measuring an error that should be closer to the "average error" over the entire experimental domain, rather than the "local" error observed for a specified set of reaction conditions.

Depending upon the experimental design chosen for a given study, DOE uses all the available information to construct a mathematical model of the entire surface within the experimental domain. Regression analysis is used to make sure the surface provides the best possible fit to the experimental data. The standard deviation obtained from the repeated data points is compared to the deviation between the model and the experimental data. If there are significant differences between the two numbers, there is lack of fit between the model and the data. In such cases, additional work should be done to identify the source of the lack of fit.

A major advantage of DOE is the model obtained from the experiments. Unlike OVAT, models obtained using DOE describe the entire surface within the experimental domain. To be most useful, care must be taken in defining the experimental domain to avoid regions in which the mechanism is known or believed to change. In addition, the range of each variable needs to be selected to avoid enclosing an experimental domain that is too large. For example, a temperature range of 30 °C for a study is much more likely to produce a model that is accurate over most of the experimental domain than a temperature range of 100 °C.

If there is concern about the size of the experimental domain, additional studies can be performed. Once the general location of a desired response has been identified, the experimental domain can be reduced to provide a more accurate response within that region. The size of the region should be selected such that it is larger than that described by the controls utilized in production. For example, if the temperature controls in a plant can only control the temperature of the reactor within ±5 °C of the set point, the experimental domain that is studied should extend beyond the ±5 °C range in order to provide information on the consequences of an improperly calibrated temperature sensor. A DOE study within this region can more precisely identify the most desirable region, provide data necessary to assess the critical parameters in the reaction, and help demonstrate that a process will be "under control" when scaled up. A knowledge of the shape of the response surface in the region to be used during manufacture of bulk chemicals can alert chemists and management to processes that are operating "on the edge of failure" and allows informed decisions during scale up and manufacturing. Processes developed using DOE are often more robust since modifications to reaction parameters are made prior to an unanticipated failure.

The calculations performed using DOE utilize matrix algebra to build the mathematical model. Although the methods are well known to mathematicians and statisticians, they are tedious and uninteresting to most organic chemists. Consequently, a good software package that simplifies data entry and analysis and presents the results in a clear, concise fashion is recommended. Numerous programs with similar capabilities are currently available. Among the better-known programs are ECHIP®, JMP®, Minitab®, MODDE®, and Design-Expert®.

Experimental Designs

To aid in understanding the following examples, brief descriptions of some of the different experimental designs are provided below. More detailed descriptions of a number of designs can be found in Deming and Morgan's book (*1*).

Factorial Design

As described previously, in the two-level factorial design experiment, each variable is set to one of two levels in any trial. The high level (+) is the upper limit for that variable in the study, the low level (-) is the lower limit. A complete two-level factorial design for three variables can be represented by the corners of a cube as shown by the black points labeled A in Figure 2. The area bounded by the cube is referred to as the experimental domain. Higher level factorial designs are possible. The addition of point B, the grey points at the center of each face, and the open points on the edges of the cube constitute a full three-level factorial design experiment in three variables. Extension of the mathematics to more than three variables is possible, but the representation of more than three variables in two-dimensional space is difficult. The total number of experiments required in a full two-level factorial set of experiments is 2^N, where N is the number of variables. The total number of experiments in a full three-level factorial set of experiments is 3^N. Addition of only the point in the center of the design (point B) allows for estimation of curvature in the model fit to the data and reduces the number of experiments from what is needed for the full three-level factorial design.

Fractional factorial designs are common. In those designs, the experimental points are often selected such that their distance from each other are maximized.

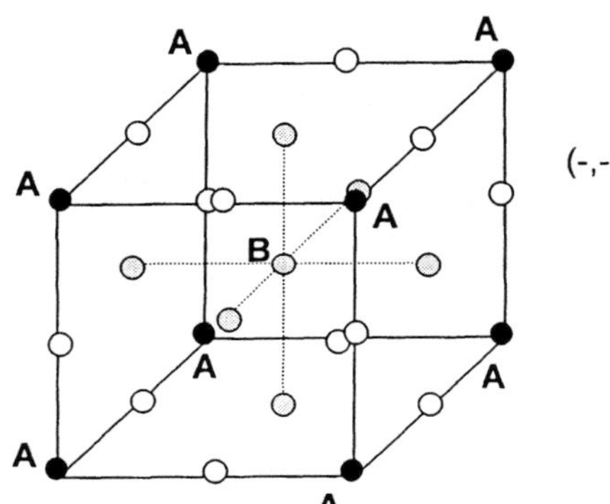

***Figure 2**. A three-level, three-factor, full factorial design.*

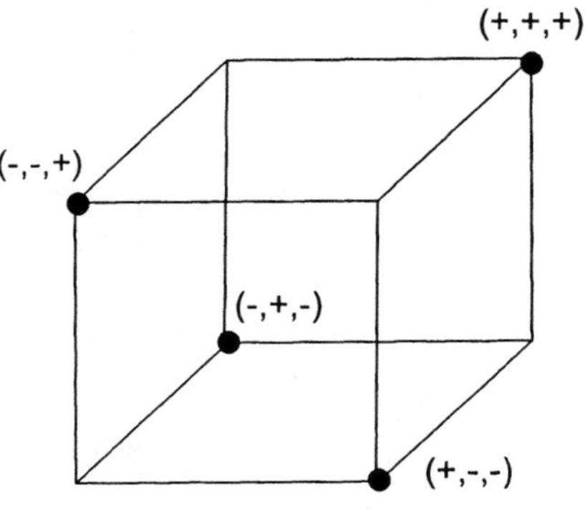

***Figure 3**. A two-level, three-factor, fractional factorial (2^{3-1}) design.*

For example, in Figure 3 the points labeled (+,+,+), (-,-,+), (+,-,-), and (-,+,-) represent four possible trials in a 2^{3-1} fractional factorial experiment (*2*). Due to the mathematics involved in solving the simultaneous equations, care must be taken in interpreting the results of such studies. In such studies individual primary effects are mathematically equivalent to specific interaction terms. As stated previously, this equivalence is referred to as "aliasing" or "confounding" of effects. In a 2^{3-1} fractional factorial study, what appear to be three primary effects may actually be two primary effects with an interaction term between those primary effects. Additional experiments may be necessary in order to determine the actual impact of each variable on the observed response and to construct a correct model.

Central Composite Design

A central composite design is similar in concept to a three level factorial design. In the central composite design, the corners (total of 2^N points) of the corresponding two-level factorial design (square points – shown as open circles on Figure 4) are augmented with a total of 2N+1 additional points that form a multi-dimensional univariate design (star points – shown as filled circles on Figure 4). Face centered central composite (ccf) designs have the points of the "star" set at the same value as the upper and lower limits of the corner points. Two additional central composite designs are possible. In a central composite circumscribed (ccc) design, the star points values are greater than the values of the corresponding corner points. In a central composite inscribed (cci) design, the experimental values of the star points are less than the values of the corresponding corner points. The central composite designs included in Figures 2 and 4 are examples of ccf designs. Figure 5 provides an example of a cci design.

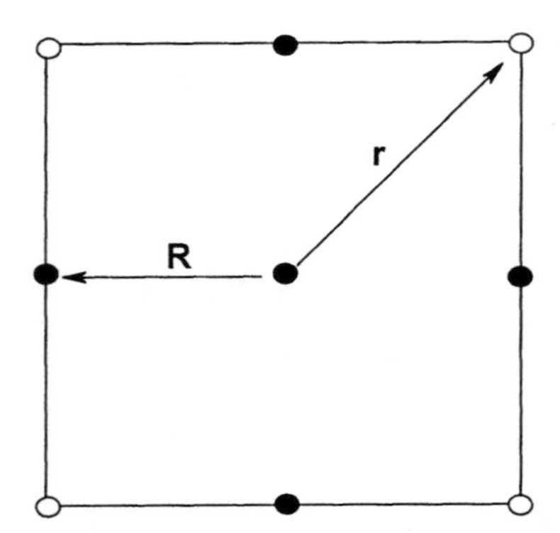

***Figure 4**. A two-factor central composite design with square points shown as circles and star points shown in black.*

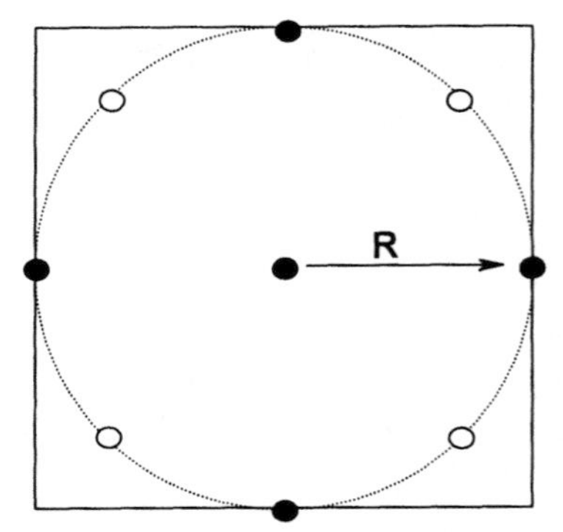

***Figure 5**. A two-factor rotatable central composite design with square points shown as open circles and star points shown in black.*

In the three level factorial design, the normalized distance (r) from the center of the design to the corners of the study is greater than the normalized distance from the center point to the midpoint of each line segment between adjacent corners (R). Consequently, the corners are weighted more than the remaining points. A special sub-class of ccc design can be made in which all experimental points on the periphery of the design are equidistant from the center (R). Such designs are rotatable central composite designs (see Figure 5). Because all points are equidistant from the center, all points are weighted equally. Despite the mathematical advantage of the rotatable central composite design, it is experimentally more difficult to set up since there are more variable levels to be considered in such studies. The variable levels may be designated as +, $\sqrt{2}/2$, 0, -$\sqrt{2}/2$, and -.

If the number of variables is two, a three-level factorial design experiment is the same as a face centered composite design as shown in Figure 4. When three or more factors are under consideration, fewer experiments are required in a central composite design than in a three-level factorial design (see Figure 2). With three factors, a minimum of 27 experiments is required in a full three-level factorial experiment. In a three-level central composite design only 15 experiments, i.e. the eight black corner points plus the five gray star points, are required. The 12 white points shown in Figure 2 are not star points and are not part of the central composite design.

Other Designs

An interesting design is the hexagonal rotatable design with center point (see Figure 6). With two factors, this design appears similar to the rotatable central composite design except the eight points around the center of the design are replaced by six points that are all equally spaced around the circumference of the circle. The coordinates of the seven experimental points can be designated as (2,0), (1, 1.732), (-1, 1.732), (-2, 0), (-1, -1.732), (1, -1.732), and (0,0). Since all points in the design are equally spaced from their three nearest neighbors, a new hexagonal rotatable design with center point can be produced using any of the non-center points by performing only three additional experiments. The design provides a great amount of flexibility if there is concern about the original design not containing the optimum experimental conditions. It is crucial to include the center point in this design.

Although numerous other designs are used they will not be discussed in this section.

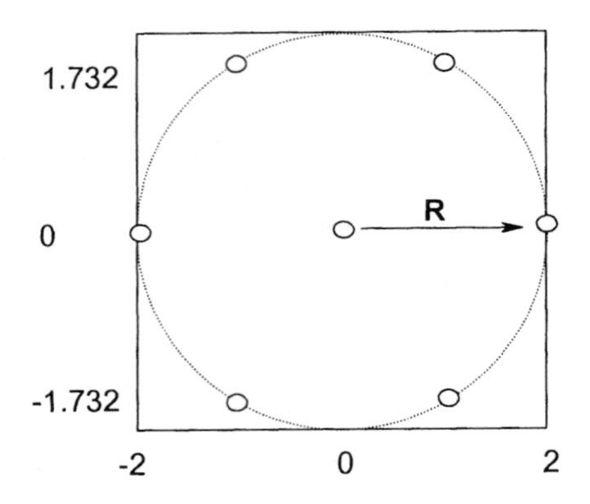

Figure 6. *A hexagonal rotatable design with center point.*

The choice of experimental design is dependent upon the model to be fit to the data as well as constraints such as time available for the study, the amount of raw materials available for the study, and other considerations. As indicated earlier, it is easy to produce experimental designs that contain more runs than are actually necessary to estimate the model coefficients for each parameter. Degrees of freedom (df) may be used to evaluate whether or not a sufficient number of runs are planned to estimate all the coefficients of the model. Degrees of freedom can be considered to be part of a bookkeeping method for comparison of the amount of information available for analysis versus the number of coefficients that can actually be estimated. For each independent run that is performed, one degree of freedom is gained. For each coefficient that is calculated during data analysis, one degree of freedom is used. One degree of freedom is always lost in calculation of the mean response. One or more df may be used in estimating the fit between the data and the model. One or more df may also be used in calculation of the replicate error. Viable experimental designs will have a net balance of greater than or equal to zero degrees of freedom after all the desired coefficients have been extracted from the raw data. It is not unusual in screening designs to have zero df for estimating lack of fit and error. Such designs are referred to as saturated. Response surface designs often have 3 to 5 df for estimating experimental error and a similar number of df for estimating lack of fit. Commercial DOE software can automatically check the df of the experimental designs and warn if there is a problem. In many cases, one may not even be aware that the degrees of freedom provided in an experimental design have been checked against the requirements of the model to be fit to the data.

Examples of DOE in Process Development

Preparation of 4-(*N,N*-Dimethylamino)acetophenone

$$\text{4-ClC}_6\text{H}_4\text{COCH}_3 + HN(CH_3)_2 \longrightarrow \text{4-(CH}_3)_2\text{NC}_6\text{H}_4\text{COCH}_3$$

Some early work utilizing DOE in process development was reported by T. Lundstedt et al. (*3*). The work is based upon a Japanese patent (*4*) describing the preparation of 4-(*N,N*-dimethylamino)acetophenone in 77% yield by heating 4-chloroacetophenone in aqueous dimethylamine. In this study, the goal was to improve the yield of product.

The variables selected for use in optimization of the process were reaction temperature and the molar ratio of dimethylamine to 4-chloroacetophenone. The experiments were run in sealed reactors. The initial design contained seven experiments (rotatable hexagonal design with center point). Six experiments were equally spaced around the center point that was set at 3 equivalents of dimethylamine and 230 °C. The unit variation in the equivalents of dimethylamine was set to 1 molar equivalent, and the unit variation in temperature was set to 20 °C. At the conclusion of this series of experiments, it was determined that the optimum conditions were not within the experimental domain. Consequently, the original study was augmented with additional experiments.

Because of the original design, only three new experiments were required to generate a new set of seven experiments that moved the new experimental domain toward the area believed to contain the optimum conditions. At the conclusion of three additional experiments, the model showed the optimum conditions to be within the new experimental domain. The optimum conditions were calculated to be 4.17 equivalents of dimethylamine at 234 °C with a predicted yield of ca. 95%. Four experiments were run using 4.22 equivalents of dimethylamine at 234.4 °C with reported yields of 89, 92, 91, and 92% of theory. After only 14 experiments, the yield of the reaction was increased by 18% with confirmation of the reproducibility of the work. In addition, the model predicted a yield of 78.5% under the reaction conditions reported in the original patent, providing additional evidence of the validity of the model.

Synthesis of 2,3:4,5-Di-*O*-isopropylidene-β-D-fructopyranose chlorosulfonate

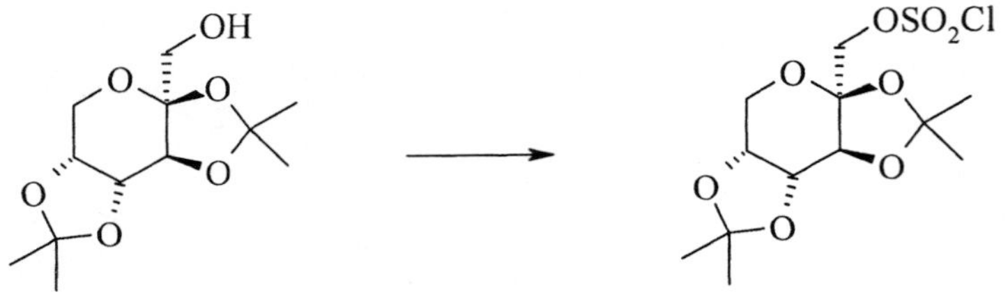

During development of 2,3:4,5-di-*O*-isopropylidene-β-D-fructopyranose chlorosulfonate, a study was designed to determine the impact of temperature and rate of addition of (2,2,7,7-tetramethyl-tetrahydro-bis[1,3]dioxolo[4,5-b;4',5'-d]pyran-3a-yl)-methanol upon the yield of crude product (Johnson & Johnson Pharmaceutical Research & Development, L.L.C., unpublished results). In the factorial design experiment, temperature was varied over a range from -10 °C to 5 °C and the rate of addition was varied such that the total addition time ranged from 2 hours to 18 hours. The data showed that there was a strong interaction between time and temperature in the reaction. At -10 °C, the yield of the chlorosulfate was about 93% regardless of the reaction time. At 5 °C, however, the yield at the end of 18-hour addition was about 86% while the 2-hour addition time produced the product in nearly 97% yield. This study established that the desired material could be produced more rapidly and at a higher temperature than had previously been considered. It is interesting to note that those working on the project had felt that the lower temperature and longer addition time would have a positive effect on the yield. DOE had led the chemists to a set of conditions that had not been considered as viable at the start of the study.

Pinner Quench Process Optimization

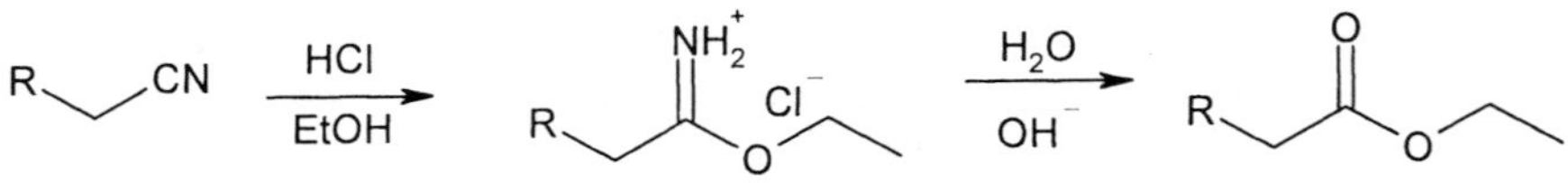

Gavin and Mojica *(6)* reported work on the optimization of the conversion of an imidate ester to an ester via hydrolysis. Factors that were felt to be important in the reaction sequence were reaction temperature (35-60 °C), the amount of water used (0.5-1.5 kg water/kg batch), and stirring speed (150-600 rpm). The initial study design was a 2^3 factorial design with four additional experiment repetitions at the center point of the design. Responses were in-situ yield and purity at several time points up to 20 hours and isolated yield. The initial work was done on a laboratory scale. Initial results showed a strong dependence of the in-situ yield on the amount of water as well as an interaction between the amount of water and the temperature. The plots of in-situ yield as a function of time showed that value to reach a maximum at about 8 hours then decrease with additional time. Conclusions drawn from the initial study included that the reaction was much faster than anticipated and that the in-situ yields were good indicators of the isolated yield. Because the process was already in a plant, additional studies on the larger scale required that the reaction conditions studied be maintained within process and regulatory limitations. On the plant scale, hydrolyses were run at 22, 35 and 45 °C with added water at levels between 0.7 and 1.2 kg/kg batch. At the conclusion of the studies, the yield of isolated product had increased by 5% to produce 71% of theory. Reaction time had decreased from 20 hours to less than 10 hours, thus producing significant increases in plant throughput.

Brederick Reaction Optimization - Synthesis of 4-(3-Pyridyl)-3*H*-imidazole

O Br N + O H_2N H → N N H N

Kirchhoff, et al (*5*) reported studies utilizing automated equipment to optimize the yield of 4-(3-pyridyl)-3*H*-imidazole from 2-bromo-1-(3-pyridyl)-1-ethanone and formamide. Initial laboratory scale work had shown that the product could be obtained in 20% yield through the use of 2 equivalents of formamide at a reaction temperature of 150 °C. Preliminary work suggested that temperature and the number of equivalents of formamide were important parameters in the reaction. The first part of the study employed a three-level factorial design experiment with five repetitions at the center point to build a quadratic model of the response surface. Temperature was varied over the range 110-170 °C and the amount of formamide was varied between 2-8 equivalents. The model indicated that the optimum conditions were outside the experimental domain included in the study. Based upon these results, a second study was conducted based upon a central composite design with five repetitions at the center. Temperature was varied between 132-188 °C and 5.2-10.8 equivalents of formamide were used. All of the above experiments were run using 0.5 g of the bromoketone and a total reaction time of 7 hours. The quadratic model constructed from the data indicated an optimum in-situ yield of product of 74 ±

3% at a reaction temperature of 160 °C using 9 equivalents of formamide. Additional work providing limitations of the equipment and the impact of those limitations on the experimental results are included in the paper.

Synthesis of Methyl 3,4,5-tri-dodecyloxybenzoate

$$\text{(HO)}_3\text{C}_6\text{H}_2\text{CO}_2\text{CH}_3 + C_{12}H_{25}Br \longrightarrow (C_{12}H_{25}O)_3\text{C}_6\text{H}_2\text{CO}_2\text{CH}_3$$

Although methyl 3,4,5-tri-dodecyloxybenzoate is not a pharmaceutical ?, the study reported by Hersmis, et al (*7*) demonstrates the power of statistical methods in process optimization. A fractional factorial experiment was designed to study the effect of starting concentration of methyl gallate (0.2-0.4 mol/L), temperature (95-115 °C), phase transfer catalyst (tetrabutylammonium bromide-Aliquat 336), amount of potassium carbonate (4-8 equivalents), amount of phase transfer catalyst (0.05-0.10 equivalent), amount of dodecylbromide (3.1-3.3 equivalents), and stirring speed (0-600 rpm). The experiments were each run using ca. 3 g of methyl gallate. The yield and selectivity of each reaction was determined using ^1H NMR analysis at 1.5 and 3 hours. A total of 19 experiments, which included 3 repetitions at the center of the design, was performed. Due to the experimental design, no confounding of primary interactions was observed and temperature, stirring speed, and the amount of potassium carbonate were found to be the most critical variables. The highest yields were obtained when all three variables were set to their high levels. At those levels nearly quantitative alkylation was observed after 1.5 hour. Interestingly, upon a 30-fold scale up, the reaction failed with only 25% of the desired product formed after 5 hours. Since these results were so different from what had been seen during the DOE work, an investigation into the cause of the poor results was conducted. The investigation revealed that a different lot of potassium carbonate had been used in the scale up and that it differed significantly in particle size from the previously used base. Grinding the potassium carbonate to an average particle size of 125 μm or smaller resulted in restoration of the previously observed yields. Thus the scale up of the reaction revealed the previously unidentified variable of potassium carbonate particle size. Without the information provided by the statistical study, it could have been more difficult to determine the cause of the scale up problem. In addition, a total of less than 30 experiments was required to select an appropriate solvent, determine the most critical of seven variables, scale the reaction, and troubleshoot the scale up. The anticipated impact of changes in specific reaction parameters was verified on the larger scale.

Suzuki Coupling Reaction

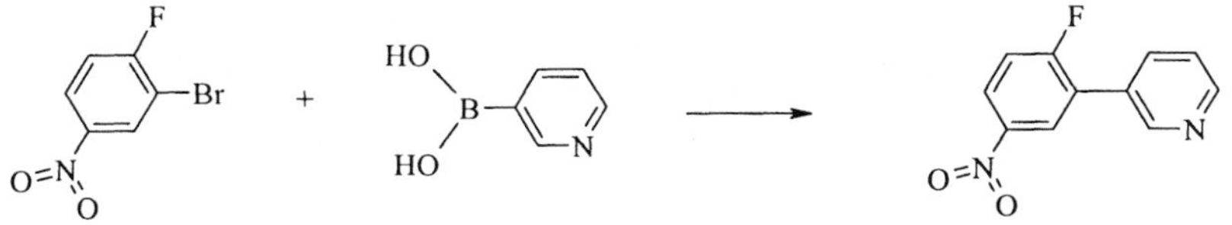

This example demonstrates problems in the interpretation of data that may be encountered even when DOE is used. The coupling of 3-bromo-4-fluoronitrobenzene with 3-pyridineboronic acid using tetrakis-(triphenylphosphine) palladium(0) in an automated system was reported by Armitage, et al (*8*). The study was a blocked four-factor fractional factorial design with two repetitions at the center point of each block. Reactions were run using 250 mg of the bromide. The factors that were studied were total solvent volume (1.55-7 mL), amount of water (0.9-2.5 mL), amount of boronic acid (1.0-1.3 equivalent), and the amount of catalyst (0.5-3 mol %). Since this reaction is known to be air-sensitive, the experiments were divided into two equivalent sets of runs (two blocks of experiments). One block was run in each of two reactors that were configured differently. The data from the two blocks were used to evaluate the performance of the two different reactor configurations. The conversion of bromide to the coupled product was evaluated after 15 hours at reflux. The study showed a specific reactor configuration to be clearly superior. Poor reproducibility was observed in one of the reactor configurations while 100% conversion in numerous experiments was observed in the second. Analysis of the complete set of data from all of the 20 experiments shows that catalyst loading and reactor configuration are statistically significant. In addition, there is an interaction between the reactor configuration and the catalyst loading. The authors interpreted the results to indicate that at high concentration, a high catalyst load can produce a robust and reliable process.

Interestingly, if only the 10 data points produced in a single reactor configuration are considered, each set of data shows no statistically significant effect from any factor. In the first piece of equipment, the lack of statistical significance is due to the high value of the standard deviation (SD = 33.58) from the replicate experiments. In the second piece of equipment the lack of statistical significance is due to the fact that there was no standard deviation (SD = 0) from the replicates.

In either of the two analyses a change in the way the response was determined may have had an impact on the conclusions. More meaningful data on the effect of all factors could possibly have been obtained if the reactions had been run for a shorter period of time. Both reactor configurations provided the opportunity to run a maximum of ten simultaneous reactions. Due to the reactor configuration, only two replicates could be performed in a given block without increasing the number of runs. While four replicates are satisfactory for the total study, a greater number of replicates in each of the smaller blocks may have provided more meaningful statistics for a single reactor configuration.

Hydrogenation of 4-Nitroacetophenone

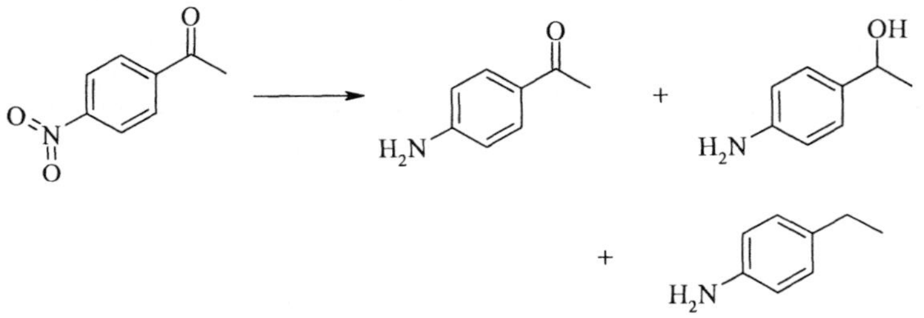

Hydrogenation of 4-nitroacetophenone can produce mixtures containing 4-aminoacetophenone, 4-(1-hydroxyethyl)aniline, and 4-ethylaniline. Hawkins and Makowski (*9*) reported the use of DOE to? optimize production of each compound independently. Screening experiments using eight catalysts, with and without methansulfonic acid in ethanol at 30 °C and 50 psi for 22 hours showed that 10% Pd/C was the best catalyst for the production of 4-(1-hydroxyethyl)aniline or 4-ethylaniline, while 5% Pd/$CaCO_3$/Pb was the best catalyst for production of 4-aminoacetophenone. With 5% Pd/$CaCO_3$/Pb as catalyst, a 2^3 factorial design experiment with center point was run using catalyst loading (5-20%), pressure (25-100 psi), and temperature (25-60 °C) as design variables. With 10% Pd/C a 2^4 factorial design with center point that included the same variables plus the amount of methanesulfonic acid (0-1.5 eq) was run. All experiments were run for 22 hours using 500 mg of substrate in 5 mL of ethanol at a stirring rate of 390 rpm. The data show that each of the three products can be preferentially formed in yields of greater than or equal to 95% depending upon the specific catalyst and reaction conditions. The best conditions are provided in Table 1.

Table 1. Best conditions for production of specific compounds by hydrogenation of 4-nitroacetophenone.

Compound	*Catalyst (Loading)*	*Pressure*	*Temp.*	*MsOH*
4-aminoaceto-phenone	5% Pd/$CaCO_3$/Pb (5%)	25 psi	25 °C	-
4-(1-hydroxyethyl)-aniline	10% Pd/C (5%)	100 psi	60 °C	0 eq.
4-ethylaniline	10% Pd/C (20%)	100 psi	25 °C	0 eq.

A total of 44 experiments was performed over the course of this study. Sixteen experiments were used to screen eight catalysts. Once the two preferred catalysts were identified, an additional 28 experiments were used to define the best conditions for production of the three reduction products.

Hydrolysis of *N*-Trifluoroacetyl-(S)-*tert*-leucine-*N*-methylamide

Although the use of DOE can reduce the amount of work necessary to reach a defined goal, the combination of DOE and automation can lead to a rapid increase in the number of experiments performed. Care should be taken in the planning of studies to clearly define the goal(s) of the work. Without clearly defined goals the utility of the method and convenience of the automation can become driving forces in a study. Rosso et al (*10*) described work on the optimization of hydrolysis of *N*-trifluoroacetyl-(S)-*tert*-leucine-*N*-methylamide. This study included the screening of 10 different bases in seven solvents, screening of three different counter-cations, and finally optimization of the amount of water, the substrate concentration and the equivalents of base used in the reaction

The original screening study used methylamine, potassium bicarbonate, potassium carbonate, ethanolamine, lithium hydride, dibasic potassium phosphate, potassium *t*-butoxide, DBU, potassium methoxide and *N,N*-diisopropylethylamine as representative bases. Solvents included in the screen were water, tetrahydrofuran, acetonitrile, ethyl acetate, methanol, ethanol, and 2-propanol. A total of 66 experiments was performed using 50 μmole (12 mg) of substrate, 2 equivalents of base, and 0.2 mL of solvent in each reaction. All reactions were heated at 45 °C for 3 hours and the area percent of the desired product was used as the response. Automated equipment was used to prepare each reaction in the study. In order to select the best solvent, the responses of all reactions were tabulated and the mean value of the response across all reactions using the same solvent was calculated. The mean value of the response for a given base was also calculated across all solvents in order to select the best base. Although DBU was found to be the best base, hydroxide was selected on the basis of lower cost.

In the second "screening study," three separate 2^3 factorial design experiments with two repetitions at the center point were run to determine the effect of temperature (30-60 °C), concentration (0.28-0.84 M), and amount of base (1-3 equivalents) on yield. Lithium hydroxide, sodium hydroxide, and potassium hydroxide were the bases used in the hydrolysis. The results from the 30 experiments in this portion of the study showed that concentration was statistically the most significant variable on yield. Lower concentrations of substrate gave higher yields. Temperature and equivalents of base were less significant. The cation was not important; therefore sodium hydroxide was selected as the base. The data also suggested the possibility of an interaction between the concentration and the equivalents of base.

The final optimization experiment used a two-factor central composite design with four repetitions at the center point. Each corner point was repeated twice giving a total of 16 runs. The concentration was varied between 0.18 M and 0.42 M and the amount of base used was between 1.075 and 1.925 equivalents. Fitting the data to a quadratic model showed only the concentration term and the square of the concentration term to be statistically significant. The yield was optimal when about 17 mL of water were used per gram of substrate. An approximately 140 fold scale up using the optimum concentration, 1.25 equivalents of sodium hydroxide at 45 °C, and a 3 hour reaction time resulted in production of a 95.6% HPLC yield of the desired product. A total of 113 reactions was run over the course of this study.

Synthesis of 17α-methyl-11β-arylestradiol

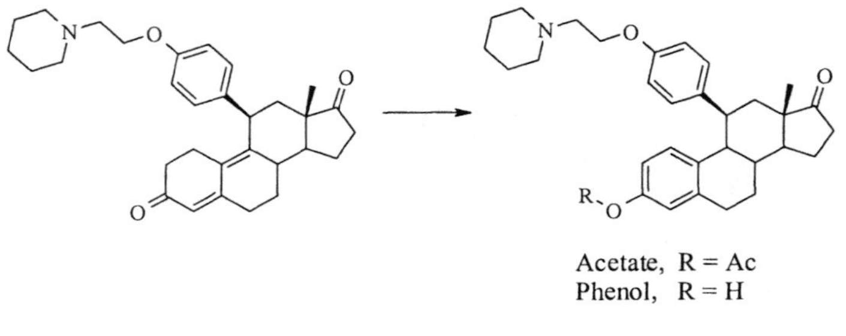

Acetate, R = Ac
Phenol, R = H

If automated equipment is used to perform a DOE study, it is important to be aware of the limitations of the equipment used in the study and their impact on the overall sequence of reactions. Larkin et al (*11*) reported a study in which the A ring of 13-methyl-11-[4-(2-piperidin-1-yl-ethoxy)-phenyl]-1,6,7,8,11,12,13,14,15,16-decahydro-2H-cyclopenta[a]phenanthrene-3,17-dione was aromatized using acetyl bromide and acetic anhydride. The three factors included in this study were the amount of acetic anhydride (0-3 equivalents), the amount of acetyl bromide (1 – 4 equivalents) and temperature (0 – 30 °C). The response was the relative amount of the acetate produced in the reaction at 5 hours. Although the experimental design was not provided, it is assumed to be a central composite design with replicates. The data were fit to a quadratic model with interactions. The data showed a maximum in the response when 1.5 equivalents of acetic anhydride, 4 equivalents of acetyl bromide and a temperature of 13-17 °C were used. In the final process, the acetate was hydrolyzed to the phenol and that compound was isolated as its hydrochloride salt. The aromatization conditions that gave the highest yield of the salt were 1 equivalent of acetic anhydride, 2.5 equivalents of acetyl bromide and a temperature of 21-24 °C. The discrepancy between the two sets of conditions is attributed to optimization of the yield of acetate in the study versus isolation of the crystalline hydrochloride salt after work up, hydrolysis, salt formation and crystallization. It was felt that the large excess of acetyl bromide used in the first step caused lower isolated yields in the reaction sequence because it interfered with the later steps. The results from the DOE were useful since they established an area in the experimental domain where the yield of the desired acetate was greater than 90%. That area was defined by the use of ≤3 equivalents of acetic anhydride, 2-4 equivalents of acetyl bromide and a reaction temperature of 15-30 °C. The best overall conditions for the reaction sequence were found to be near the center of this area on the response surface. The hydrochloride salt was isolated in 82.7% yield and 98.7% purity versus 77.3% yield and 98.1% purity before the study. The authors concluded that the study, which consisted of 19 experiments, resulted in a significant improvement in terms of cost, safety, hygiene, and environmental implications over the previously used conditions.

Conclusion

The application of DOE in organic synthesis during process development is a powerful tool. Care must be exercised at the start of any study to define its goals. Experimental designs appropriate to defined goals are important to success. During data analysis, one must be aware of the limitations of the original design to make sure aliasing has not caused misinterpretation of the data. The information gained through a well-planned and well-executed DOE study is greater than that obtained through the same number of experiments preformed using OVAT experimentation.

There are numerous software packages available for use in DOE. While the details of each package may vary, a function of the software is to expedite the use and interpretation of DOE data. A basic understanding of statistics and experimental design is helpful, but not necessarily a requirement. Most DOE software can be used without a full understanding of the process being used by the software to solve the simultaneous equations and produce the mathematical model of the system. It is useful to have access to a statistician who understands DOE in order to address questions concerning any study designs, lack of fit, aliasing, or other issues.

The design and execution of DOE studies does not require robotics in order to be effective. The combination of DOE and automation can simplify the generation, collection and interpretation of large amounts of data within a short time frame. Automated reactors may improve the reproducibility of a study, but the conclusions drawn from a well-executed study should be the same with or without their use. Automated systems can reduce much of the redundancy involved in DOE work but they are no panacea for the thought that is involved in the selection of appropriate goals, experimental parameters and evaluation tools while studying a reaction.

The pharmaceutical industry faces a continuing emphasis on the rapid development of safe, efficient processes for the production of new drugs. DOE can play an important role as a tool that provides for recovery of the maximum amount of information from a limited number of experiments. As such, its role within chemical development must continue to grow as more chemists discover its utility.

References

1. Deming, S.N.; Morgan, S.L. *Experimental design: a chemometric approach*; Data Handling in Science and Technology Vol. 11; Elsevier: NY, 1993.
2. Designations for partial factorial design experiments commonly are provided in the form L^{k-p} where L represents the number of levels used in the study for each factor, k is the number of factors in the study, and p is used to determine the fraction of the full factorial in the study. To determine the fractional size, one must calculate the value of $(1/2)^p$. The total number of experiments (excluding any repetitions) is determined by calculating the value of $L^{(k-p)}$. The value of the designation is that it provides much information in a minimal amount of space. For example, 3^{4-1}

indicates that four factors were studied at three levels in a one-half fractional design requiring 3^3 or 27 experiments.
3. Lunstedt, T.; Thorén, P.; Carlson, R. *Acta Chem. Scand.* **1984** *B38, 717-19.*
4. Yamamoto, K.; Nitta, K.; Yagi, N.; Imazato, Y.; Oisu, J.; *Jpn. Pat.* 54132542, Mitsui Toatso Chemical Inc. (*Chem. Abs.* 92, 163712 (1979)).
5. Kirchhoff, E.W.; Anderson, D.R.; Zhang, S.; Cassidy, C.S.; and Flavin, M.T. *Organic Process Research & Development* **2001** *5(1), 50-53.*
6. Gavin, D.J.; Mojica, C.A. *Organic Process Research and Development,* **2001** *5(6), 659-664.*
7. Hersmis, M.C.; Spiering, A.J.H.; Waterval, R.J.M.; Meuldijk, J.; Vekemans, J.A.J.M.; Hulshof, L.A. *Organic Process Research & Development* **2001** *5(1), 54-60.*
8. Armitage, M.A.; Smith, G.E.; Veal, K.T. *Organic Process Research & Development* **1999** *3(3), 189-195.*
9. Hawkins, J.M.; Makowski, T.W. *Organic Process Research & Development,* **2001** *5(3), 328-330 .*
10. Rosso, V.W.; Pazdan, J.L.; Venit, J.J. *Organic Process Research & Development* **2001** *5(3), 294-298.*
11. Larkin, J.P.; Wehrey, C.; Boffelli, P.; Lagraulet, H.; Lemaitre, G.; Nedelec, A.; and Prat, D. *Organic Process Research & Development* **2002** *6, 20-27.*

Chapter 7

Process Development for the MMP Inhibitor AG3433

Jayaram K. Srirangam[1], Ming Guo[1], Shu Yu[1], Alan W. Grubbs[1], James Saenz[1], Steven L. Bender[2], Judith G. Deal[2], Kuenshan S. Lee[1], Jason Liou[1], Robert Szendroi[1], James Faust[1], and Kim Albizati[1]

[1]Chemical Research and Development and [2]Medicinal Chemistry, Pfizer Global Research and Development, La Jolla/Agouron Pharmaceuticals, Inc., 3565 General Atomics Court, San Diego, CA 92121

AG3433 is a selective and potent inhibitor of the Matrix Metalloproteinase (MMP) class of enzymes. The development of an efficient process for the large-scale production of the drug substance is presented. Various synthetic routes to this 3-substituted *N*-aryl pyrrole are discussed. An efficient method for the *N*-arylation of substituted pyrroles is reported.

Introduction

The Matrix Metalloproteinases (MMPs) are a family of zinc-containing, calcium-dependent enzymes documented to be involved in the proteolysis of the basement membrane and other extracellular matrix (ECM) components, and they appear to play an essential role in angiogenesis, tumor growth, and metastasis. Inhibition of these enzymes have been proposed as a possible means of controlling growth and metastasis of tumors while exhibiting a low toxicity profile compared to existing therapies. Several reviews on the design of MMP inhibitors have appeared in the literature, which included among others, Bay 12-9566 (Bayer), Marimastat (BB-2516), and Prinomastat (Agouron) (*1*) (Figure 1). AG3433 was developed as a 3rd generation inhibitor of MMPs specifically designed to avoid the joint pains believed to have originated from the inhibition of MMP-1 (also called collagenase) (*2*).

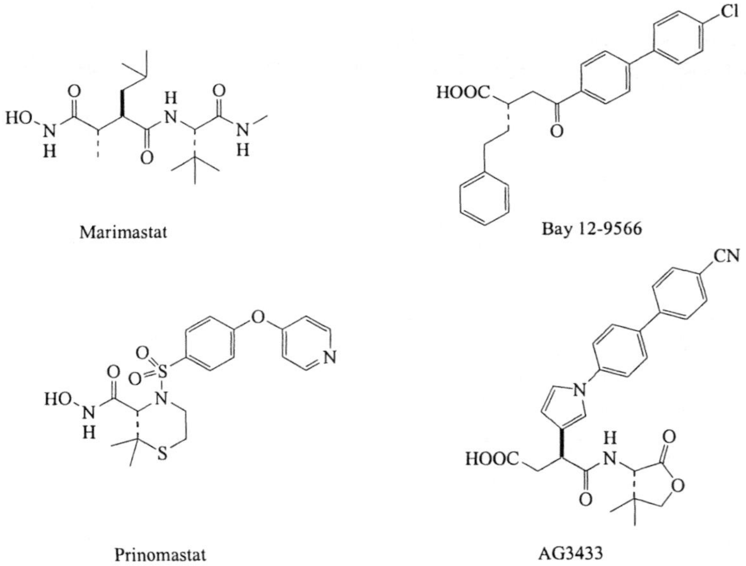

Marimastat

Bay 12-9566

Prinomastat

AG3433

***Figure 1**. MMP inhibitors*

Synthesis of AG3433

AG3433 may be considered as a substituted succinic acid mono amide. The early synthesis of AG3433 (Scheme 1) began with 2,5-dimethoxy tetrahydrofuran-3-carboxaldehyde **1**. The one carbon homologation of the aldehyde was carried out *via* dithiane chemistry. The ester **4** was condensed with 4'-amino-4-cyano biphenyl (**5**) to provide the *N*-aryl pyrrole acetic acid methyl ester **6**. Hydrolysis of the ester followed by reaction with the chiral auxiliary **8** resulted in the *N*-acyloxazolidinone intermediate **9**. Alkylation of **9** with benzyl bromoacetate gave the substituted succinic acid half ester **10**. Removal of the chiral auxiliary followed by coupling with aminolactone **12** resulted in the benzyl ester **13**. The benzyl group was removed by hydrogenolysis to give AG3433.

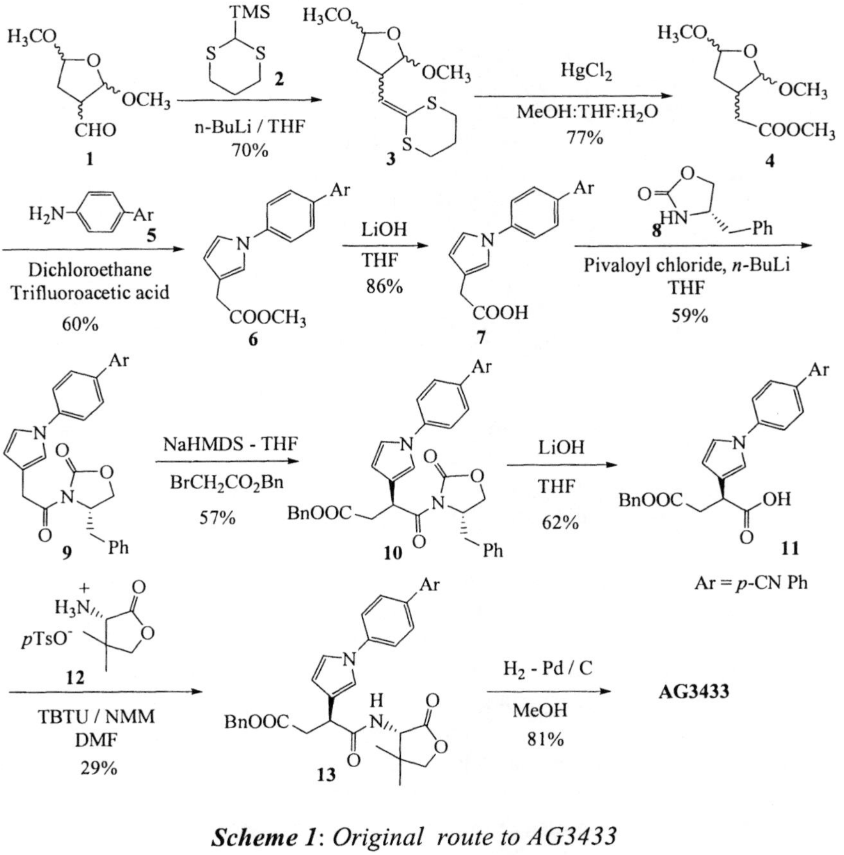

***Scheme 1**: Original route to AG3433*

The amino pantolactone required for this synthesis was obtained from the commercially available (*R*)-pantolactone **14** (*3*). The triflate **15** was converted to the azide **16** using tetra-*n*-butyl ammonium azide. The azide was found to be stable with an onset temperature of 200 °C making it safe and scalable. Hydrogenation of the azide was carried out in presence of *p*-TsOH to obtain the required aminolactone as the *p*-toluene sulfonate salt (Scheme 2).

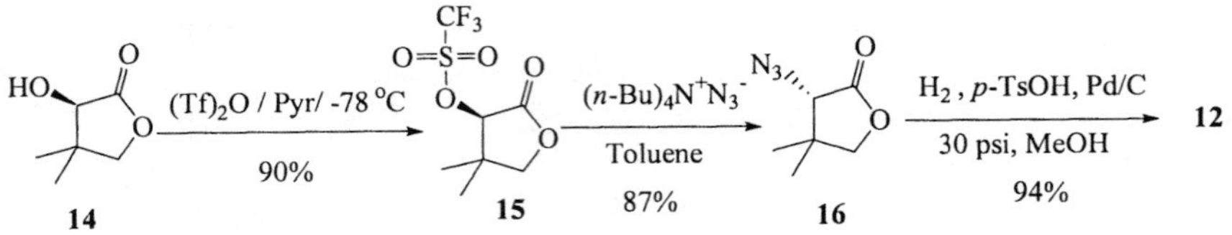

Scheme 2**: Synthesis of Aminopantolactone **12

The cyano biarylamine used in the synthesis was obtained from the commercially available 4-cyano biphenyl, **17** (Scheme 3). Nitration of **17** in trifluoroacetic anhydride resulted in a 3:1 mixture of 4'- and 2'-nitro compounds. From the mixture, the required 4'-nitro compound was separated by crystallization from acetone-*iso*-propanol. Reduction of the nitro group was carried out with tin(II) chloride and conc. HCl to afford the required product in 90% yield.

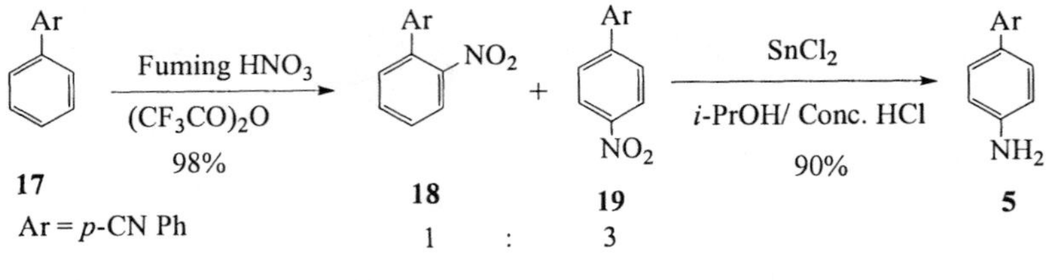

Scheme 3**: Synthesis of cyanobiarylamine **5

Early investigations towards the scale-up synthesis of AG3433 revealed several problems with the original route:

- Both the starting material **1** and 2-TMS dithiane **2** were expensive and not available in commercial quantities.
- Although the use of mercuric chloride was replaced with silver nitrate, the chemistry remains unsafe and non-scalable. Furthermore, the aryl acetic acid **7** obtained *via* this route was contaminated with inorganic impurities.
- *p*-(*p*-Cyanophenyl)aniline (**5**) is a suspected carcinogen.

- The alkylation of the *N*-acyloxazolidinone **9** suffered from a poor diastereomeric selectivity (5:1 ratio at best).
- Isolation/purification of the succinic acid mono-ester **11** was difficult.
- Coupling of acid **11** with aminolactone **12** resulted in poor yields and partial racemization.
- Hydrogenolysis of the benzyl group resulted in impurities difficult to remove from the final drug substance.

Process Improvements

The serious limitations in the synthesis of the aryl acetic acid **7**, required the development of a better alternative. Retrosynthetic analysis of the aryl acetic acid led us to compound **21** as a possible starting material. Compound **21** is prepared from the commercially available and inexpensive 2,5-dimethoxy dihydrofuran **20** by free radical addition of malonic ester. Compound **21** was obtained in commercial quantities from a vendor but the purity was <40%. Reaction of crude **21** with *p*-(*p*-cyanophenyl)aniline (**5**) afforded the aryl malonic ester **22**. This was hydrolyzed to the diacid, following a solvent exchange with dioxane to remove the methanol, and decarboxylated with hydrochloric acid at 90°C to give the aryl acetic acid in a one-pot operation (Scheme 4). The aryl acetic acid obtained in this process was >98% pure.

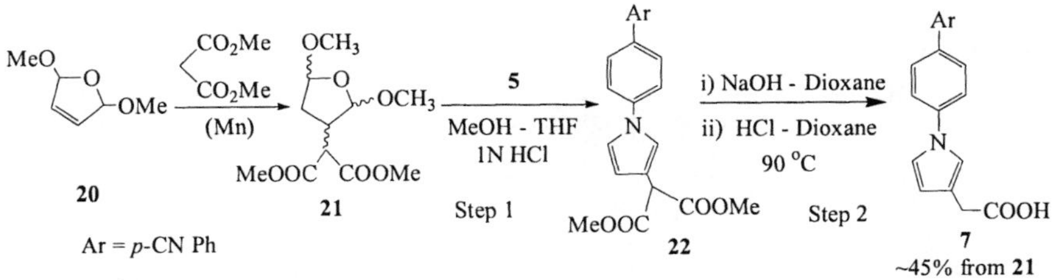

***Scheme 4**: Improved synthesis of aryl acetic acid 7*

Formation of the *N*-acyloxazolidinone **9** (Scheme 5) was optimized to occur at -40°C. Use of acetic acid in the work up resulted in a good phase split during the extraction. The unreacted acid could be removed by bicarbonate wash as it stayed in the aqueous phase as a fine dispersion during the extraction. Alkylation of the *N*-acyl oxazolidinone was studied in greater detail. Of the various electrophiles and bases tried, *t*-butyl bromoacetate/NaHMDS combination gave the best results. The desired diastereomer **10a** was obtained in a 96:4 ratio and did not require further purification.

The chiral auxiliary was removed with LiOH and the resulting acid was isolated as the dicyclohexylamine salt **11a**. The salt **11a** was then slurried in acetonitrile resulting in an increased purity to >95%. The salt was used directly in the next coupling reaction. The coupling was attempted with various peptide coupling agents. HATU worked best with yields >90%. The reaction was carried out at RT in DMF without epimerization at the α-carbon. The solid product was isolated by slowly adding the reaction solution to a large excess of water (20vol/vol) with vigorous agitation. Various acids and solvents were studied for the removal of the *t*-butyl ester. It was discovered that concentrated HCl in acetonitrile accomplished the conversion in less than 1h. The final product, AG3433 was recrystallized from ethanol/water to obtain the drug substance in >99% purity with a diastereomeric content <0.15%.

New Route Development

Although an enabling route to AG3433 was developed, it still suffered from some major drawbacks. The malonyl tetrahydrofuran **21** was expensive and only available in <40% purity, which required the use of silica gel for the purification of **22** after the coupling to remove the impurities. Use of the chiral auxiliary on large scale was not economical. An alternate approach was sought that would use commercially available inexpensive starting materials. Thus, 3-substitution on *N*-aryl pyrroles was considered as an alternate approach to AG3433.

Generally, the 2-position in pyrrole, is more reactive towards electrophilic substitution reactions(*4*). Attempted Friedel-Crafts acylation of *N*-phenyl pyrrole resulted in a mixture of products with no substitution on pyrrole. Use of nitromethane as a solvent, which has been reported to assist in homogeneity and affect the rate of Friedel-Crafts acylations (*5*), gave a mixture of 2- and 3-acylated products in the ratio of 3:1. Thus it was necessary to block the 2-position to force reactions at the 3-position of *N*- phenyl pyrrole (*6*). The Vilsmeier reagent, which has found its use in such acylations on unsubstituted pyrroles (*7*) was chosen for this purpose. Reaction of *N*-phenyl pyrrole with oxalyl chloride and DMF followed by *in situ* acylation with ethyl oxalyl chloride and aluminum chloride gave the desired 4-acylated product **24** (Scheme 6). The use of nitromethane as a solvent helped in slowing down the reaction rate and resulted in a homogeneous reaction mixture. The reaction worked best with 4.4 equivalents of aluminum chloride. Under the optimum conditions only 5 - 10% of the 5-acylated product was formed and it was easily removed by reslurry of the crude product in MTBE. The aldehyde **24** was easily decarbonylated by heating with Pd/C. Wolff-Kishner reduction of the keto ester gave the acetic acid derivative **27**.

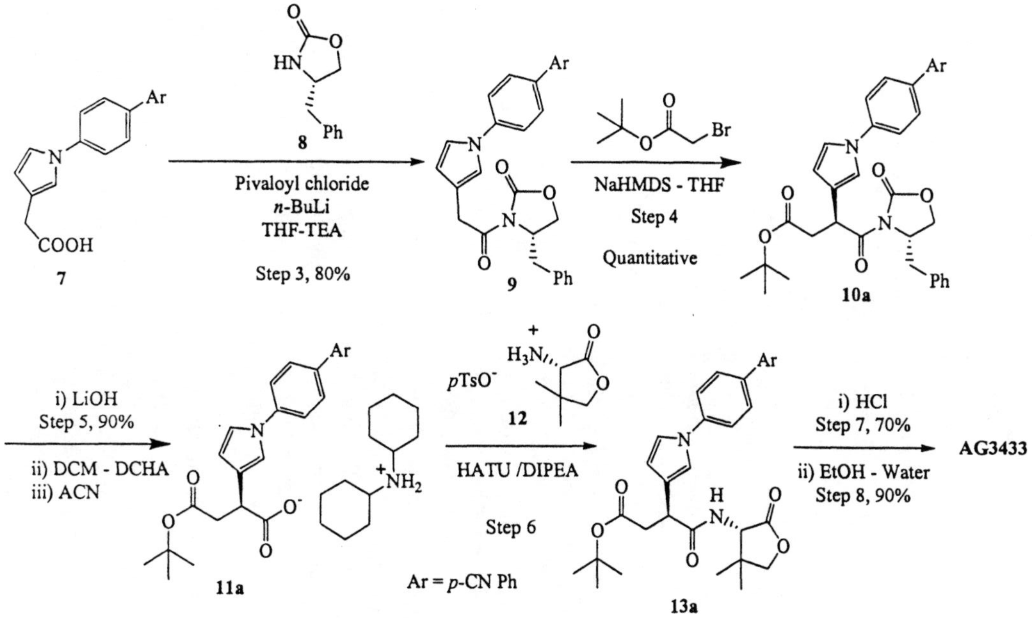

Pivaloyl chloride
n-BuLi
THF-TEA
Step 3, 80%

NaHMDS - THF
Step 4
Quantitative

i) LiOH
Step 5, 90%
ii) DCM - DCHA
iii) ACN

$pTsO^-$ H_3N^+
12

HATU /DIPEA
Step 6

Ar = p-CN Ph

i) HCl
Step 7, 70%
ii) EtOH - Water
Step 8, 90%

AG3433

7 8 9 10a 11a 13a

***Scheme 5**: Process chemistry synthesis of AG3433*

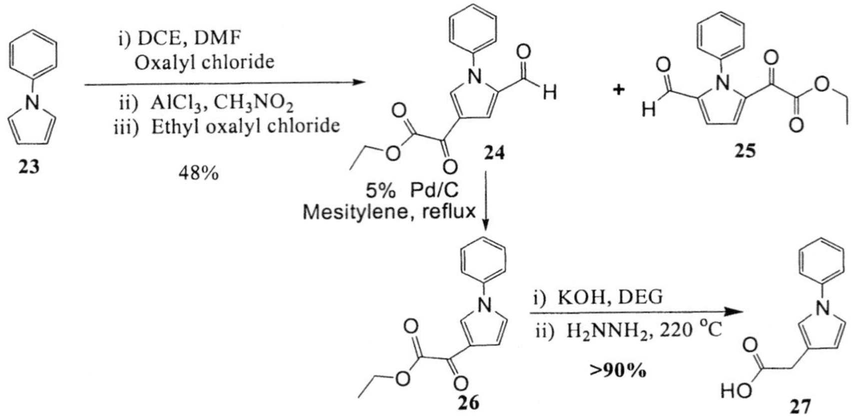

Scheme 6: 3-Substituion on N-phenyl pyrrole

With the above methodology in hand, a similar strategy was attempted for the synthesis of the aryl acetic acid **7** (Scheme 7). The aniline **5** was treated with 2,5-dimethoxytetrahydrofuran in toluene and acetic acid to get the *N*-aryl pyrrole **28**. The Vilsmeier/Friedel-Crafts acylation of **28** followed by decarbonylation afforded the keto ester **30**. However, the reduction of the keto ester proved to be difficult. Most of the general methods employed for the reduction of ketones gave a mixture of products. However, a two step process involving the formation of the thioketal **31** followed by desulfurization with Ni proved to be successful. Although, the standard Wolff-Kishner conditions could not be employed in the system due to the susceptibility of the nitrile to hydrolysis, a modified Wolff-Kishner reduction (*8*) proved to be fruitful.

Although the above route was successful in providing an easy access to the aryl acetic acid **7**, it was still plagued by the use of *p*-(*p*-cyanophenyl)aniline (**5**), which is a suspected carcinogen. Thus it was decided to use a different *p*-substituted aniline and later extend it *via* a Pd mediated process. Commercially available and inexpensive *p*-bromo aniline was the compound of first choice. Condensation with dimethoxy tetrahydrofuran followed by Vilsmeier/Friedel-Crafts reaction sequence resulted in the formation of the required aldehyde **35**. However, the decarbonylation of this aldehyde proved to be extremely difficult. With fewer options available, we decided to convert the bromide to the boronate ester **37** by reacting with bis(pinacolato)diboron **36**. To our surprise, the boronate ester **37** underwent decarbonylation readily with Pd/alumina to give the keto ester **38**. The keto group was reduced by treatment with either Ra-Ni or Pd-

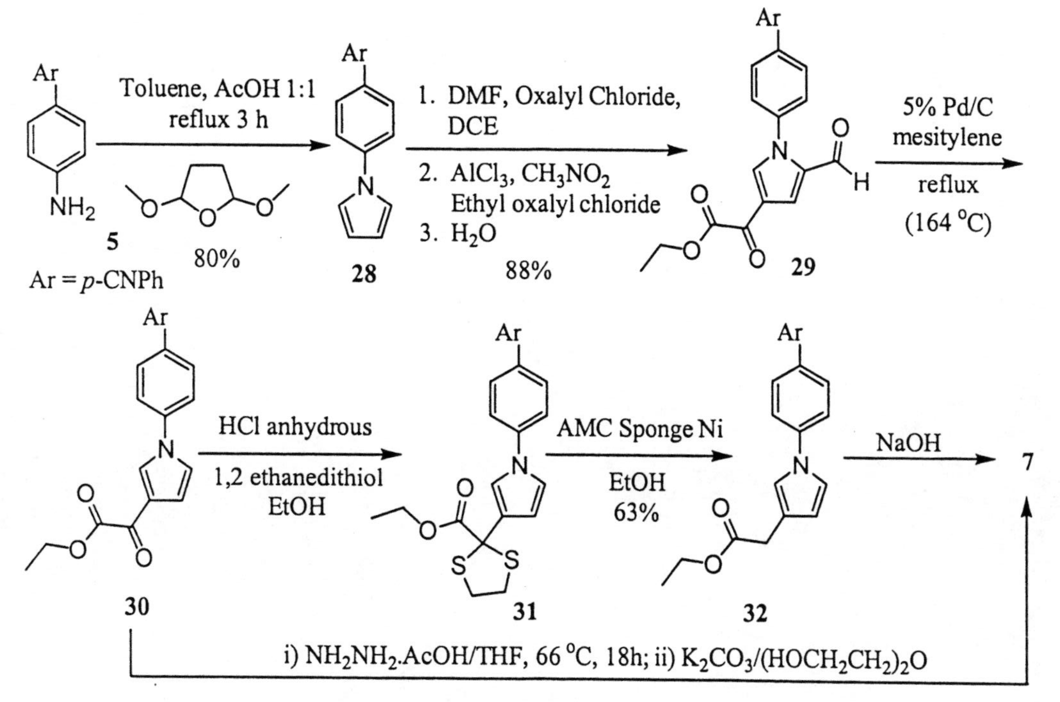

***Scheme** 7: Synthesis of aryl acetic acid 7 via the Vilsmeier-Haack/Friedel Craft's acylation*

C and the resulting boronate coupled with *p*-bromo benzonitrile **40** to give to the required aryl acetic acid ethyl ester **32** (Scheme 8).

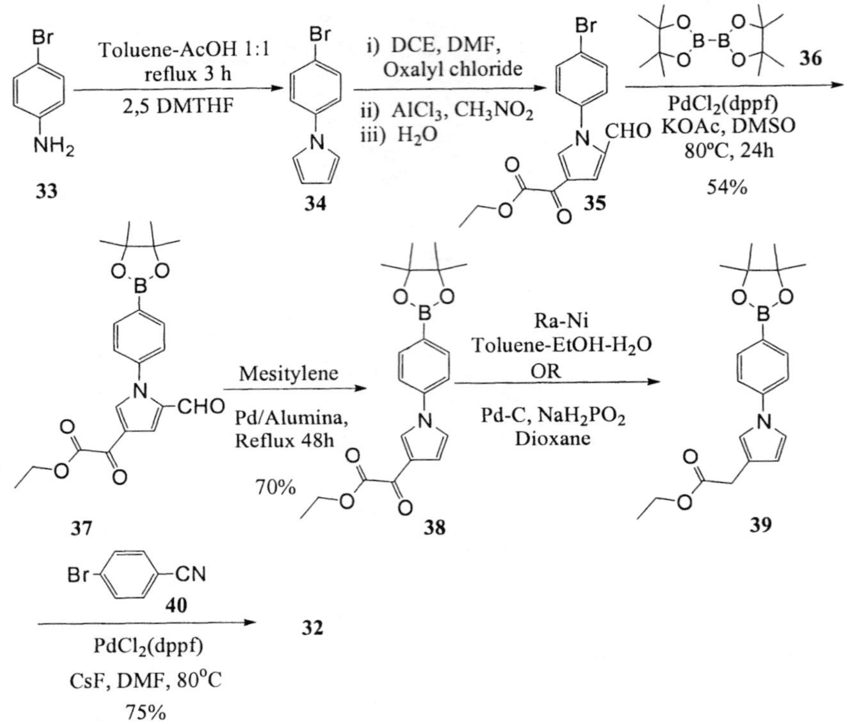

Br NH$_2$ **33**; Toluene-AcOH 1:1, reflux 3 h, 2,5 DMTHF → **34**; i) DCE, DMF, Oxalyl chloride, ii) $AlCl_3$, CH_3NO_2, iii) H_2O → **35** (CHO); **36**, $PdCl_2(dppf)$, KOAc, DMSO, 80°C, 24h, 54% → **37** (CHO); Mesitylene, Pd/Alumina, Reflux 48h, 70% → **38**; Ra-Ni, Toluene-EtOH-H_2O OR Pd-C, NaH_2PO_2, Dioxane → **39**; Br–C$_6$H$_4$–CN **40**, $PdCl_2(dppf)$, CsF, DMF, 80°C, 75% → **32**

***Scheme 8**: Synthesis of iaryl acetic acid ester **32** from p-bromo aniline*

The above route also proved to be economically less viable, therefore, we decided to use the commercially available *p*-amino phenol to carry out a similar set of reactions. *p*-Amino phenol was condensed with dimethoxy tetrahydrofuran to obtain the *N*-aryl pyrrole **41**. The phenol **42** was converted to the acetate and that was subjected to the Vilsmeier/Friedel-Crafts acylation conditions to obtain the aldehyde **43**. Decarbonylation of the aldehyde followed by Wolff-Kishner reduction and esterification resulted in the formation of compound **45**. Similar results were obtained starting with *p*-anisidine. In this case demethylation was carried out using boron tribromide. The phenolic group was converted to the triflate and coupled with *p*-cyano phenylboronic acid **47** (*9*) to provide the desired product **6** (Scheme 9).

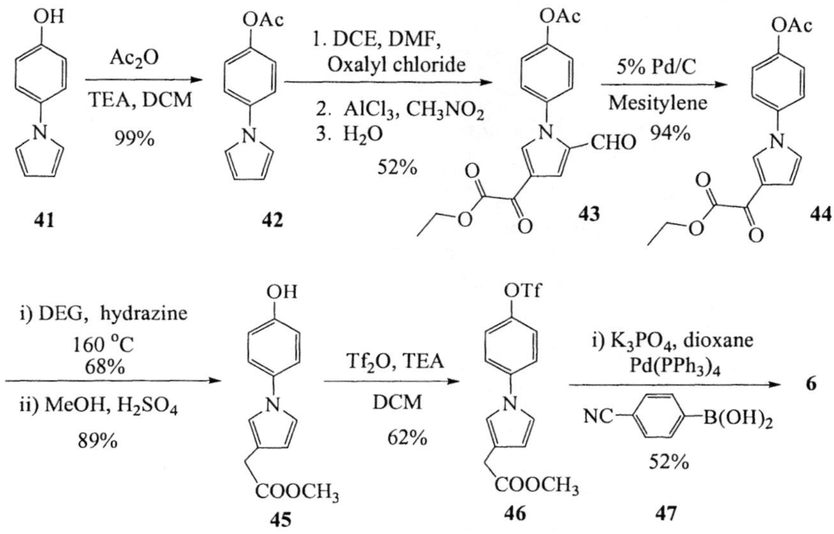

Scheme 9: *Synthesis of iarylacetic acid from p-amino phenol*

N-Arylation of Pyrroles

An alternate pathway to the arylacetic acid **7** would involve *N*-arylation of suitably substituted pyrrole. However, *N*-arylation of pyrroles has been difficult to achieve(*10*). We were able to form the C-N(*sp2*) bonds efficiently by the *N*-arylation of electron deficient pyrroles using aryl boronic acids with $Cu(OAc)_2$ as the catalyst (*11*).

This methodology was successfully applied to the synthesis of compound **29,** a pivotal intermediate in the synthesis of the aryl acetic acid **7**. As indicated in Scheme 10, keto ester **29** was prepared in excellent yield from the corresponding boronic acid **49** by coupling with the pyrrole **50** (*13*).

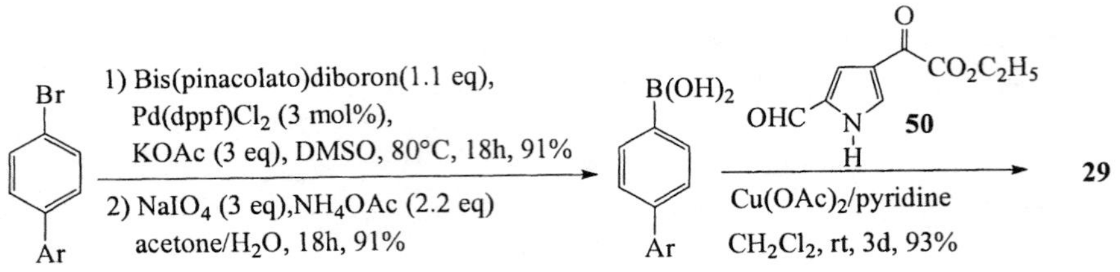

Br

1) Bis(pinacolato)diboron(1.1 eq), $Pd(dppf)Cl_2$ (3 mol%), KOAc (3 eq), DMSO, 80°C, 18h, 91%

2) $NaIO_4$ (3 eq),NH_4OAc (2.2 eq) acetone/H_2O, 18h, 91%

Ar

48 Ar = *p*-CN-Ph

$B(OH)_2$

Ar

49

O

$CO_2C_2H_5$

OHC N H **50**

$Cu(OAc)_2$/pyridine

CH_2Cl_2, rt, 3d, 93%

29

***Scheme 10**: Synthesis of intermediate **29** through the N-arylation methodology*

Due to the high cost of the nitrile **48**, we developed an alternate route to synthesize **29** that involved the use of the inexpensive 4,4'-dibromo biphenyl **51** (Scheme 11). Compound **51** was converted to the boronic acid **52** by treatment with *n*-BuLi and trimethyl borate. The boronic acid was coupled with **50** and the resulting halide was converted to the nitrile **29** by treatment with $Zn(CN)_2$ (*13*).

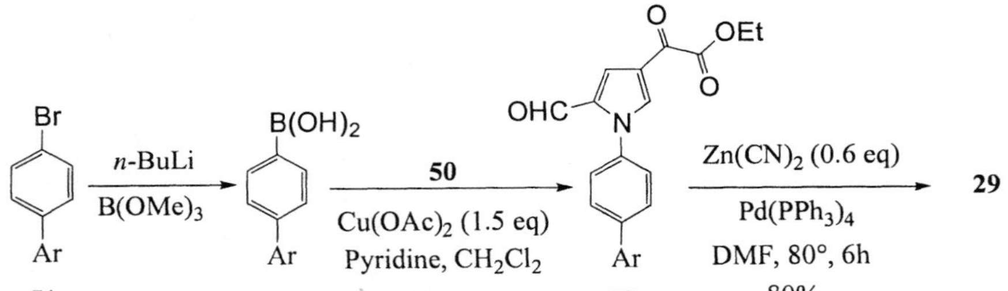

Br

n-BuLi

$B(OMe)_3$

Ar

51

$B(OH)_2$

Ar

52

50

$Cu(OAc)_2$ (1.5 eq)

Pyridine, CH_2Cl_2

74%

O

OEt

O

OHC N

Ar

53

$Zn(CN)_2$ (0.6 eq)

$Pd(PPh_3)_4$

DMF, 80°, 6h

80%

29

Ar = *p*-Br Ph

Scheme 11**: Synthesis of intermediate **29** from 4,4'-dibromo biphenyl **59

Thus, the early development work on the synthesis of AG3433 resulted in an efficient process for the large scale production of the drug substance as well as several alternative routes to the key intermediate **7** from commercially available starting materials.

Acknowledgements

We wish to thank Drs. Srinivasan Babu, Van Martin, John Roberts (Hoffmann-La Roche), Phil Roberts, and Ms. Bridgette Craig-Woods for their assistance during the course of these studies.

References

1. Skiles, J. W., Monovich, L. G.; Jeng, A. Y. In *Annual Reports in Medicinal Chemistry*; Doherty, A. M., Ed.; Academic Press: San Diego, CA, **2000**, *35*, 167-176.
2. Deal, J. G.; Bender, S. L.; Chong, W. K. M.; Duvadie, R. K.; Caldwell, A. M.; Li, L.; McTigue, M. A.; Wickersham, J. A.; Appelt, K.; Almassy, R. J.; Shalinsky, D. R.; Daniels, R. G.; McDermott, C. R.; Brekken, J.; Margosiak, S. A.; Kumpf, R. A.; Abreo, M. A.; Burke, B. J.; Register, J. A.; Dagostino, E. F.; Vanderpool, D. L.; Santos, O. *217th ACS National Meeting*, Anaheim, CA, March 21-25 **1999**, MEDI-197.
3. Freskos, J. *Syn. Comm.* **1994**, *24*, 557-563.
4. Jackson, A. H. In *Pyrroles Part One*; Ed. Jones, A. R.; John Wiley and Sons: New York, **1990**, 295-303.
5. Gore, P. H. In "*Friedel-Crafts and Related Reactions*", Ed. Olah, G. A.; Interscience: New York, **1964**, *Vol. III Part I*, 1-381.
6. Rokach, J.; Hamel, P.; Kakushima, M. *Tetrahedron Lett.* **1981**, *22(49)*, 4901-4904.
7. (a) Barker, P.; Gendler, P.; Rapoport, H. *J. Org. Chem.* **1978**, *43*, 4849-4853; (b) Belanger, P. *Tetrahedron Lett.* **1979**, 2504-2508; (c) Anderson, H. J.; Loader, C. E.; Foster, A. *Can. J. Chem.* **1980**, *58*, 2527-2530.
8. Paquette, L. A.; Varadarajan, A.; Bay, E. *J. Am. Chem. Soc.* **1984**, *106*, 6702 – 6708.
9. Ornstein, P. L.; Zimmerman, D. M.; Arnold, M. B.; Bleisch, T. J.; Cantrell, B.; Simon, R.; Zarrinmayeh, H.; Baker, S. R.; Gates, M.; Tizzano, J. P.; Bleakman, D., *J. Med. Chem.* **2000,** *43 (23),* 4354 - 4358.
10. (a) Mann, G.; Hartwig, J. F.; Driver, M. S.; Fernandez-Rivas, C. *J. Am. Chem. Soc.* **1998**, *120*, 827-828. (b) Hartwig, J. F.; Kawatsura, M.; Hauck, S. I.; Shaughnessy, K. H.; Alcazar-Roman, L. M. *J. Org. Chem.* **1999**, *64*, 5575-5580.
11. Yu, S.; Saenz, J.; Srirangam, J. K. *J. Org. Chem.* **2002,** *67*, 1699-1702.
12. Demopoulos, B. J.; Anderson, H. J.; Loader, C. E.; Faber, K. *Can. J. Chem.* **1983**, *61*, 2415-2422.
13. Tschaen, D. M.; Desmond, R.; King, A. O.; Fortin, M .C.; Pipik, B.; King, S.; Verhoeven, T. R. *Synth. Commun.* **1994**, *24(6)*, 887-890.

Chapter 8

Dihydro-7-benzofurancarboxylic Acid: An Intermediate in the Synthesis of the Enterokinetic Agent R108512

Bert Willemsens[1], Alex Copmans[1], Dirk Beerens[1], Stef Leurs[1], Dirk de Smaele[1], Max Rey[2], and Silke Farkas[2]

[1]Chemical Process Research Department, Janssen Pharmaceutica, Turnhoutse weg 30, 2340 Beerse, Belgium
[2]Chemical Process Research Department, Cilag Ltd., Hochstrasse 201/209, 8201 Schaffausen, Switzerland

A facile and scalable method for the synthesis of 4-amino-5-chloro-2,3-dihydro-7-benzofurancarboxylic acid, an intermediate in the synthesis of the enterokinetic agent R108512, has been developed. The key step in the synthesis is a zinc mediated ring closure of methyl 4-(acetylamino)-3-bromo-2-(2-bromoethoxy)-5-chlorobenzoate. The ring closure can be achieved without preliminary activation of the zinc on condition that the oxygen content in the reaction mixture is lower than 0.5%. The new process eliminates hazardous chemicals (ethyleneoxide) and low temperature reactions (-70°C, *n*-BuLi)

Introduction

Constipation is the most common gastrointestinal complaint, with more than two million patient visits in the United States per year, and probably an even larger number of individuals trying to remedy the symptom by self-medication.[*1*] Slow transit constipation, a particular form of constipation for which no clear underlying cause has been defined, is characterized by a lack of massive bowel movements in the large intestine. This type of movement is responsible for the transport of the colonic content along the large bowel. In essence, slow transit constipation can be regarded as a result of a motor dysfunction, i.e. a lack of prokinetic activitity in the large intestine. It is well documented that stimulation of serotonin 5-HT_4 receptors in the gut increases the motility throughout the gastrointestinal tract.[*2,3*] Therefore it is believed that gastrointestinal prokinetic agents have an important role to play in the treatment of slow transit constipation. Prucalopride,[*4*] a highly potent and selective serotonin 5-HT_4 agonist, clearly demonstrates efficacy in clinical studies.[*5-10*] Prucalopride is isolated either as the hydrochloric acid salt R093877 or as the succinate R108512 (Figure 1).

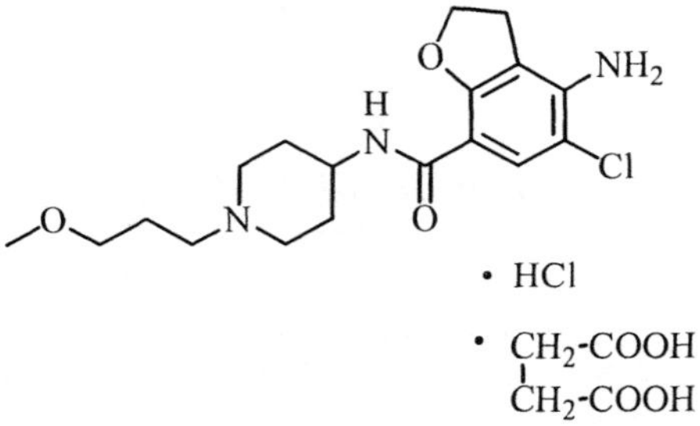

• HCl R093877

• CH_2-COOH / CH_2-COOH R108512

Figure 1*. R108512*

An important intermediate in the synthesis of prucalopride is 4-amino-5-chloro-2,3-dihydro-7-benzofurancarboxylic acid **1**, which after coupling with 1-(3-methoxypropyl)-4-piperidinamine **2,** gives the anilide **3** yielding R093877 or R108512 after salt formation (Scheme 1).

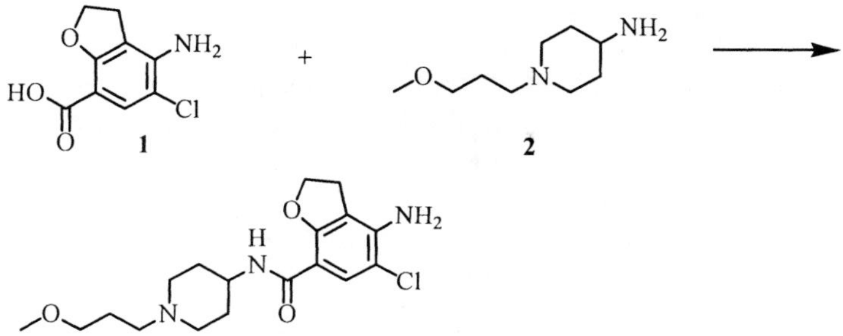

Scheme 1. *Penultimates*

The medicinal chemistry route to prepare the anilide **3** , as outlined in Scheme 2, follows the procedure of Marburg and Tolman.[*11*] Though this route was applied to prepare the early toxicology and clinical batches, the use of ethylene oxide, benzene, *N*-chloro- and *N*-bromosuccinimide together with the low temperatures required for the *n*-BuLi reactions prevented the introduction of this method into our chemical production plant.

Besides these inconveniences, chlorination of 4-amino-2,3-dihydrobenzofuran **4** to the 5-chloro derivative **5** is not regioselective and a chromatographic purification was necessary to remove the 7-chloro isomer **6**.

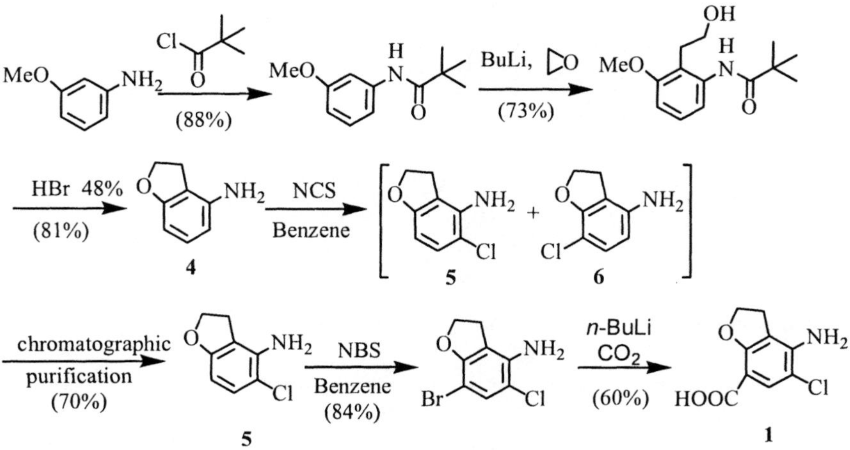

Scheme 2*. Medicinal Chemistry route*

Alternative Routes

In 1993 Bedeschi et al.[*12*] published the synthesis of the dihydrobenzofurancarboxylic acid **1** using a palladium catalysed coupling of trimethylsilylacetylene with methyl 4-(acetylamino)-3-iodo-2-hydroxy-5-chlorobenzoate. Most of the alternatives we investigated were performed prior to this publication and moreover trimethylsilylacetylene is not a reagent of choice to introduce in a production plant due to it's high price and low atom efficiency.

Several alternatives to the synthesis of **1** have been investigated and besides the final method, two of them merit consideration in some detail.

A three step route to compound **5** is possible via the application of Marburg and Tolman's method to the commercially available 2-chloro-5-methoxyaniline **7** as outlined in Scheme 3. Starting with this aniline eliminates the regioselectivity problems encountered during the chlorination. Ring closure of the hydroxyethylated compound **8** however resulted in the formation of the dihydroindole **9** rather than the desired dihydrofuran **5**.

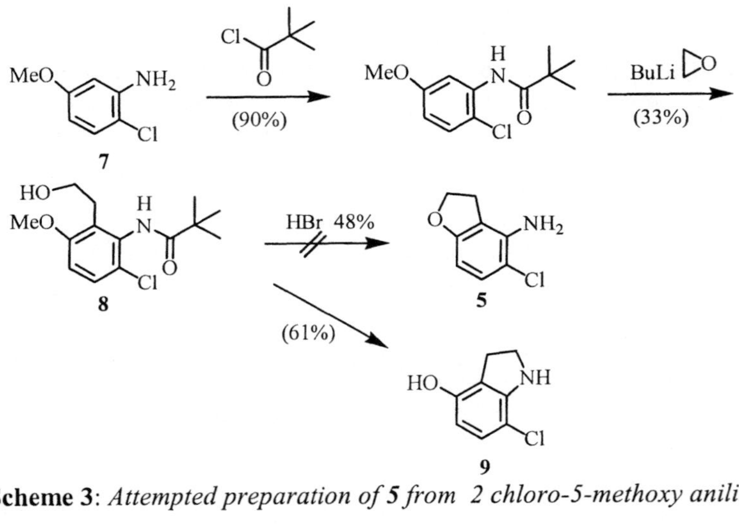

Scheme 3: *Attempted preparation of 5 from 2 chloro-5-methoxy aniline*

Another approach to the synthesis of the dihydrobenzofurancarboxylic acid **1** is outlined in Scheme 4. The key reaction in this alternative is the Houben-Hoesch reaction on methyl-4-amino-2-methoxy-5-chloro benzoate **10.** Houben-Hoesch reaction on anilines, giving exclusively ortho amino ketones has been described by Sugasawa.[*13*] Under these reaction conditions, the ether function is demethylated to provide the keto-phenol **11**. Compound **11** can then be cyclized under very mild conditions to the furanone **12** which, after reduction and subsequent dehydration gives the benzofuran derivative **13**. Hydrogenation of this benzofuran to the dihydrobenzofuran **14** was accomplished with rhodium as catalyst. Though the reduction proceeded quantitatively, the isolated **14** contained about 4% of dechlorinated product. As earlier investigations indicated, this impurity most probably could be reduced to an acceptable level leading to a scalable method in our pilot plant. For the final production method however the use of borontrichloride would create significant amounts of boron-containing waste and had to be eliminated. Therefore, in parallel to the Houben-Hoesch approach another alternative was investigated.

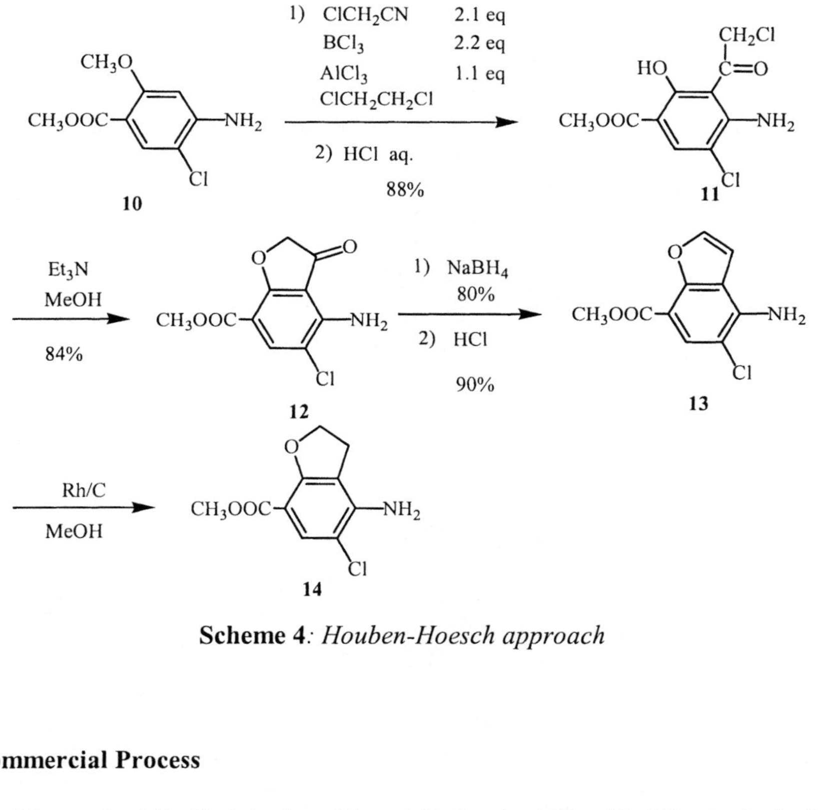

1) $ClCH_2CN$ 2.1 eq; BCl_3 2.2 eq; $AlCl_3$ 1.1 eq; $ClCH_2CH_2Cl$; 2) HCl aq.; 88% — **10** → **11**

Et_3N, MeOH, 84% — **11** → **12**

1) $NaBH_4$ 80%; 2) HCl 90% — **12** → **13**

Rh/C, MeOH — **13** → **14**

Scheme 4: *Houben-Hoesch approach*

Commercial Process

The method finally introduced in production is outlined in Scheme 5. In the following section the development will be described in more detail.

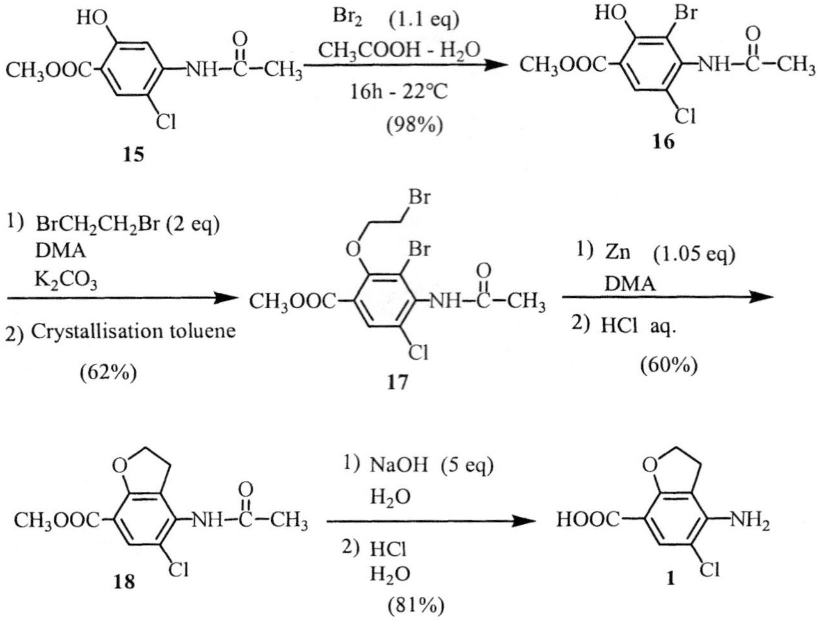

Scheme 5: *Commercial process*

Bromination Step

Bromination of the commercially available salicylic acid derivative **15** was first attempted in pure acetic acid and two equivalents of bromine to obtain an acceptable conversion. As shown in Table 1, lower excess of bromine and longer reaction times or elevated temperatures lead to the formation of the deacetylated product as an impurity. When switching to an aqueous solvent system much better conversions were obtained and the amount of bromine could be reduced to 1.1 equivalent. As illustrated in Table 2, bromination in aqueous acetic acid gave almost complete conversion and the product could be filtered directly from the reaction mixture. Bromination in water gave incomplete conversion.

Table I: Bromination in anhydrous acetic acid
Results after 16 and 20 hours are on isolated compound, other results are in process controls.

eq. Br_2	*Temp. °C*	*Time*	*LC wt%*		*Deacetylated product.*
			Product	*starting material*	
2	20 - 30	16 h	94	2	
	20 - 45	3 h	98	1.5	
1.8	20 – 45	3 h	93	3	
1.5	20 – 45	1 h	89	8	
		16 h	69	2	21
1.25	20 – 30	5 h	82	14	
		20 h	87	8	
1.25	40 - 80	2 h	76	2	20

***Table II*: Bromination in aqueous solvent systems**
Results after 16 hours at 22-26°C.

Solvent	**LC wt% (isolated product)**		**Yield**
	product	**starting material**	
H_2O	96.5	3.3	71%
H_2O / CH_3COOH (1/1)	99.3	0.03	98%

Bromoethylation

In screening experiments bromoethylation of phenolic derivative **16** gave the best results in dimethylacetamide (DMA) as solvent and using K_2CO_3 as base. The excess of dibromoethane (10 equivalents) required further investigation. Besides being a suspected carcinogen, the large excess of dibromoethane also caused problems during the work-up of the reaction. The reaction mixture, once quenched in water, was extracted with dichloromethane (another solvent to be avoided in production plants). From this extract the solvents and excess dibromoethane were evaporated giving an oily residue, still containing some dibromoethane, from which the product precipitated after addition of water. The residual quantity of dibromoethane in the residue was critical and prevented precipitation of the product when exceeded 30% by weight. When the excess of dibromoethane was reduced to 2 equivalents these problems were solved; the work-up was simplified and made more robust. Furhtermore, instead of precipitating, the product crystallized from the reaction mixture after addition of water.

A drawback of working with 2 equivalents of dibromoethane was that the isolated product contained 4 to 8% of compound **19** (Figure 2) whereas with 10 equivalents this amount was only 2 to 3%. This problem was solved by recrystallizing the crude, wet product from toluene to give the bomoethyl ether **17** in 62% yield and a purity exceeding 98%.

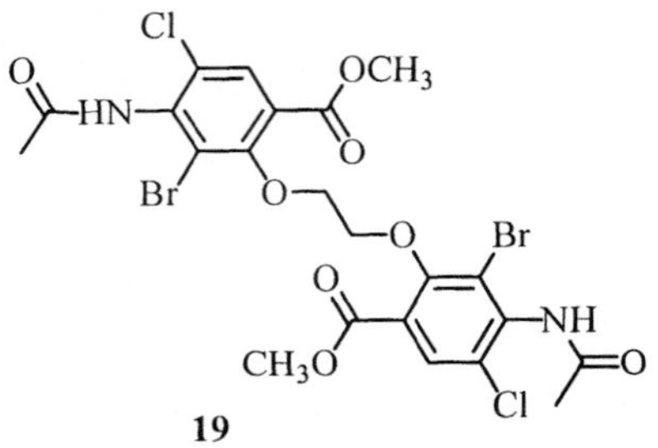

19

Figure 2: *Structure of double alkylation product*

Zinc-Mediated Ring Closure

The crucial step in this approach was the ring closure of the dibromo derivative **17** to the dihydrobenzofuran **18**. Ring closure of ortho brominated 2-haloethylphenylethers using Mg or Li to the corresponding 2,3-dihydrobenzofurans is well documented in literature.[*14-18*] These reagents however are not compatible with the other functional groups in our substrate. Organozinc reagents are known to have a high functional group compatibility.[*19*] Moreover, cross coupling reactions catalyzed by organozinc reagents are well known.[*19,20*] These data prompted us to investigate a zinc mediated ring closure of **17**. Direct insertion of zinc into an organic halide is known to be difficult [*21, 22*] and zinc needs to be **activated**.[*19, 23, 24*] In our case, initial experiments showed that the reaction proceeded even without zinc activation and that addition of transition metals, such as Pd, did not improve the conversion. These experiments were run under a nitrogen atmosphere in DMA using 1.1 to 1.3 equivalents of zinc at 40 - 80 °C. Lower conversions were obtained in tetrahydrofuran (THF). A typical conversion is shown in Table 3, the structures of the impurities are outlined in Figure 3. Use of zinc powder (<60 µm) or granular zinc (30 mesh) gave comparable conversions. On large scale however, granular zinc particles are too heavy to become equally distributed within the reaction mixture and will remain mainly in the lower part of the reactor. As this could influence the conversion, zinc powder was selected for further optimization.

Table III: Conversion of 17 to 18 in screening experiments after 22 hours
(In process controls; HPLC – relative weight percent)

Solvent	*Zn*	*Temp. (°C)*	*18*	*17*	*20*	*21*	*22*
DMA	Powder	40	64.1	1.3	7.6	2.3	12.5
		60	66.3		7	7.6	3.3
		80	55.8		7.2	5.7	13.8
	Granulate	60	65.4		8.5	12.1	1.6
THF	Powder	60	29.1	29.6	7.7	2	20

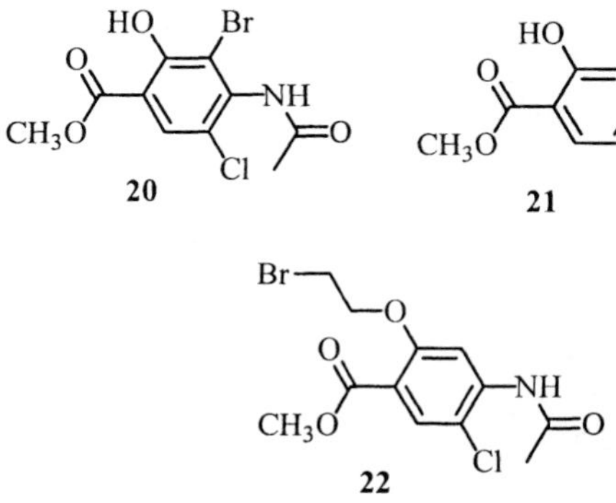

***Figure 3**: Impurities formed during the ring closure*

Numerous screening experiments were performed; in general higher reaction temperatures (80°C or higher) resulted in more side products. At lower temperatures (40 – 60°C) the start of the reaction and the exotherm involved were not reproducible. Therefore, the reaction was studied in more detail in the reaction calorimeter. An initial experiment revealed that the heat of reaction was 333kJ/mole of compound **17** undergoing ring closure. The reaction concentration used (1L DMA /mole **17**) resulted in an adiabatic temperature rise of 122°C. Thus starting the reaction at 60°C could lead to a potentially unsafe situation as the reaction temperature might rise above the boiling point of DMA in case of a cooling failure. On the other hand, it was noticed that minor amounts of oxygen (introduced in the reaction mixture during sampling) immediately stopped the reaction. This is illustrated in Figure **4**. Thus in case of a cooling failure the exotherm could be prevented by introducing air into the reaction.

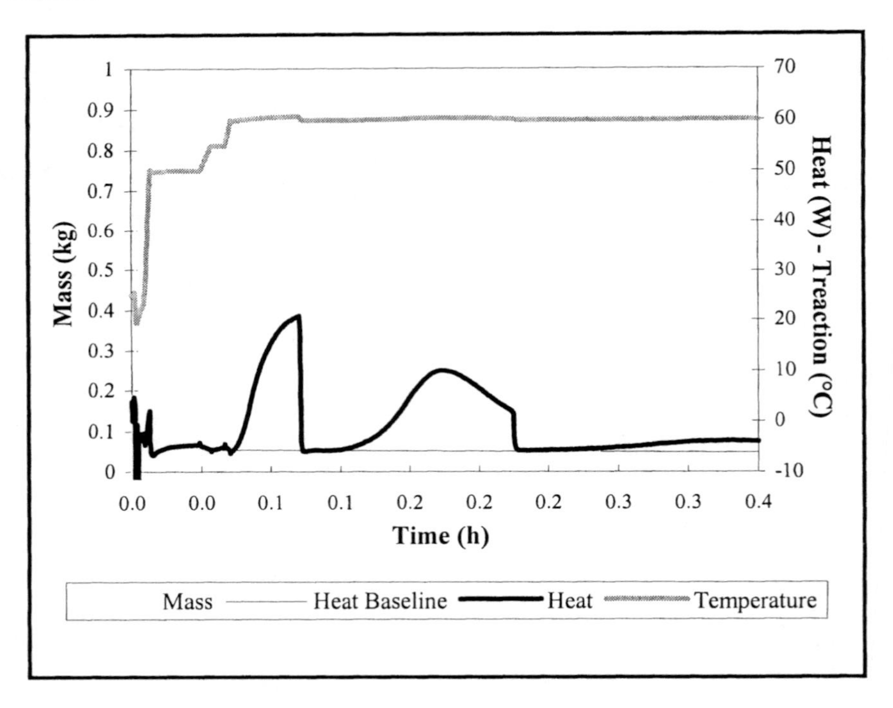

Figure 4: *Reaction inhibition by oxygen*

The role of the oxygen content in the reaction mixture at the start of the reaction was further investigated. Applying vacuum to the reaction mixture followed by flushing with argon, the oxygen content in the mixture was reduced to a level below 0.5%. As shown in Figure **5**, at this oxygen level, reaction started at 20°C. When the inert atmosphere was interrupted the reaction immediately stopped but started again as soon as the inert atmosphere was re-established.

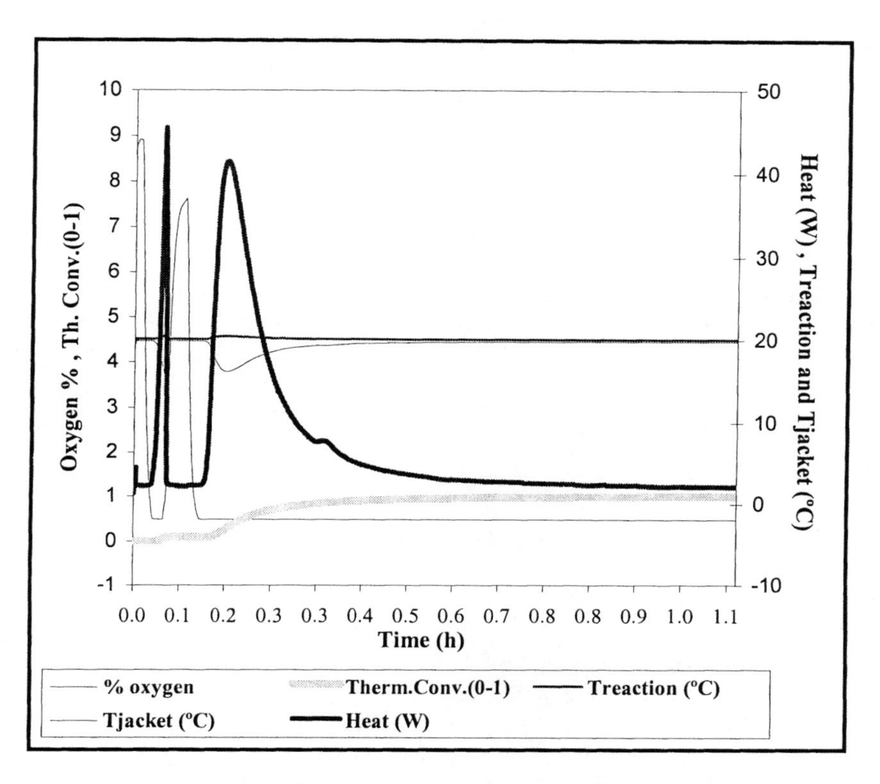

Figure 5: Effect of oxygen on the start of the reaction

This procedure afforded us a safe and scalable method for the ring closure of **17** to **18**. The product crystallized from the reaction mixture after quenching the excess zinc with hydrochloric acid and subsequent addition of water. This crystallization however was not selective; the isolated product **18** contained about 8% of **16** and 1.5% of **15**, impurities which were removed in the next step.

The method was introduced in production and several batches on an 800 to 1600 mole scale were performed successfully.

Hydrolysis

As the dihydrobenzofuran carboxylic acid **1** is a penultimate intermediate in the synthesis of R108512, it is essential that this compound doesn't contain any impurities that might impact the final API (active pharmaceutical ingridient) purity. This goal was achieved by isolating the sodium salt of compound **1** prior to acidification. Hydrolysis of crude anilido-ester **18** with NaOH afforded the sodium salt of **1**, which crystallized from the reaction mixture. The pure salt was isolated by filtration leaving the impurities in the filtrate. Subsequent acidification gave the desired acid **1** in HPLC purity of 99%.

Conclusion

The synthesis of **1** by a zinc-mediated ring closure of **17** resulted in a process applicable for a multi purpose production plant. The new procedure reduced the number of isolations from 6 to 4. It eliminated the use of the hazardous reagents, ethyleneoxide and butyllithium, low reaction temperatures and chromatographic separation. Furthermore, the overall isolated yield was improved from 18 to 29%.

Experimental section

General

Compound **15** was supplied by AMSA. Screening experiments on the zinc mediated ring closure were performed with zinc powder supplied by Merck (K16748689) and zinc granulate supplied by ACROS (30 mesh – 22261). Calorimetric experiments were performed in a RC_1 Mettler reaction calorimeter with zinc powder supplied by Umicore (the former Union Minière).

Chemical purities were determined by HPLC on a Hewlett-Packard liquid chromatograph HP1090 series II using a LichroCART 125-4 Superspher 60-RP Select B column (Merck) and UV-detection at 254 nm wavelength. Separations were achieved with a linear gradient of 1g H_3PO_4 85% in 100 mL water and 1g H_3PO_4 85 % in 100 mL acetonitrile as mobile phase at a 1.2 mL/min. flow.

^{1}H NMR spectra were measured on a Bruker AMX400 spectrometer in DMSO-d_6.

Oxygen content in the reaction mixture was measured with a “Knick Prozess O2 meter-73O2” together with an oxygen probe.

Methyl 4-(acetylamino)-2-hydroxy-3-bromo-5-chloro benzoate (16). A 5-L four necked round-bottom flask was charged successively with compound **15** (122 g, 0.5 mole), acetic acid (1 L) and water (1 L). To the heterogeneous mixture bromine (88 g, 0.55 mole) was added dropwise over a period of 1 hour at 22-26°C. The mixture was then stirred for another 16 to 20 hours at this temperature. The precipitate was filtered and washed twice with water (2 x 75 mL portions). The product was dried in vacuo at 50°C for 20 hours. Yield: 160 g

(98%). HPLC: 99.28%. ^{1}H NMR (400 MHz, DMSO – d_6) δ 2.06 (s, 3H) 3.92 (s, 3H) 7.87 (s, 1H) 10.05 (s, 1H) 11.18 (s, 1H)

Methyl 4-(acetylamino)-3-bromo-2-(2-bromoethoxy)-5-chloro benzoate (17) A 3-L four necked round-bottom flask was charged successively with compound **16** (161.5 g, 0.5 mole), potassium carbonate (69.1 g, 0.5 mole) and dimethylacetamide (1 L). The heterogeneous mixture was heated to 50°C and stirred for one hour. 1,2-dibromoethane (188 g, 1 mole) was added and the mixture was stirred for another 5 hours at 50°C. Warm (45 – 60°C) water (0.75 L) was added. The resulting homogeneous mixture started to crystallise 5 to 10 minutes after the addition of water. The mixture was then stirred for 1 hour at 50°C after which another portion (0.25L) of warm water was added. The mixture was then cooled to 22°C and stirred for another 5 to 20 hours. The precipitate was filtered and washed with water (0.1 L). The wet product was transferred into a 3-L four necked round-bottom flask equipped with a Dean Stark trap. Toluene (2 L) and filter aid (dicalite, 25 g) was added and the mixture was refluxed, while distilling off the water, until the internal temperature reached 110°C. The mixture was filtered hot leaving impurity **19** on the filter. The filtrate was crystallised while cooling to 15°C for 16 hours. the product was dried in vacuo at 55°C for 16 hours. Yield: 132 g (62%)

HPLC: 98.5%. ^{1}H NMR (400 MHz, DMSO – d_6) δ 2.07 (s, 3H) 3.80 (dd, *J*=5.92, 5.20 Hz, 2H) 3.86 (s, 3H) 4.29 (*J*=5.53 Hz, 2H) 7.88 (s, 1H) 10.08 (s, 1H)

Methyl 4-acetylamino-5-chloro-2,3-dihydro-7-benzofurancarboxylate (18) A Mettler RC_1 calorimeter, equipped with a 'Knick Prozess-O_2 meter $73O_2$ and oxygen probe, was charged successively with compound **17** (214.7 g, 0.5 mole), dimethylacetamide (0.5 L) and zinc powder (64.3 g, 0.525 mole). The oxygen content in the reaction mixture was reduced to <0.5% by inertisation with nitrogen. the mixture was stirred at 20°C. After about 0.5 to 2 hours an increase of the Q_r signal indicated that reaction had started. The mixture was then stirred for another 16 to 20 hours at 50°C. A solution of concentrated hydrochloric acid (54 mL, 0.6 mole) in water (150 mL) and 2-propanol (90 ml) was added dropwise while heating the mixture further to 70°C. After stirring for an additional hour at 70°C, water (125 mol) was added dropwise and stirred for 15 minutes. After cooling to 60°C another portion of water (195 mL) was added dropwise. The mixture was stirred for one hour at 60°C during which crystallisation started. Then the mixture was cooled to 22°C and stirred for 18 hours. The precipitate was filtered, washed with water (90 mL) and dried in vacuo at 50°C for 20 hours. Yield: 85 g (63%).

HPLC: 88.3 % (6.6 % of compound **15**). ^{1}H NMR (400 MHz, DMSO – d_6) δ 2.07 (s, 3H) 3.05 (t, *J*=8.79 Hz, 2H) 3.78 (s,3H) 4.66 (t, *J*=8.8 Hz, 2H) 7.62 (s,1H) 9.91 s,1H)

4-Amino-5-chloro-2,3-dihydro-7-benzofuran carboxylic acid (1) A 1-L four necked round-bottom flask was charged successively with compound **18** (67.42 g, 0.25 mole), water (337 mL) and propyleneglycol monomethylether (67.5 mL). A 50wt% solution of sodiumhydroxide in water (100 g, 1.25 mole) was added. The resulting suspension was heated slowly to 90°C and stirred for 16 hours at this temperature. The mixture was cooled with ice water to 5°C and stirred for another 2 hours. The precipitate (sodium salt of **1**) was filtered and washed twice with water (2 10 mL portions). The wet product was transferred again into the 1-L flask and water (500 mL) and propyleneglycol monomethylether (87.5 mL) were added. The mixture was heated to 80°C at which temperature it became homogeneous. Hydrochloric acid 11N (23 mL) was added. During the addition the acid **1** started to precipitate. The mixture was cooled to 10°C and stirred for another 2 hours. The precipitate was filtered and washed twice with water (2x15 mL portions). The product was dried in vacuo at 50°C for 16 hours. Yield: 43.5 g (81.4%, 93.4% calculated on the net amount of **18** charged into the reaction).HPLC: 99.0 %. ^{1}H NMR (400 MHz, DMSO – d_6) δ 2.96 (t, *J*=8.85 Hz, 2H) 4.58 (t, *J*=8.90 Hz, 2H) 5.98 (s, 2H) 7.41 (s, 1H) 12.05 (s, 1H)

Acknowledgements

We wish to acknowledge many Janssen and Cilag colleagues for their support in this project. In particular we thank Veronique Cerpentier, An Vandendriessche, Leen Schellekens, Tony Nelen, Dirk Daemen, Karel Laenen, Lieven Van Ermengem, Jean-Paul Bosmans, Rolf Hänseler, R. Weibel and Hansueli Bichsel.

References

1. Schiller, L. R. Chronic constipation: Pathogenesis, Diagnosis, Treatment. In *Evolving concepts in gastrointestinal motility.* Champion, M. C., Orr, W. C., Eds.; Blackwell Science Ltd.: Oxford, 1996; pp 221-250.
2. Tam, F. S.; Hillier, K.; Bruce, K. T. *Br. J. Pharmacol.* **1994**, *113*, 143-150.
3. McLean, P. G.; Coupar, I. M. *Br. J. Pharmacol.*, **1996**, *117*, 238-239.
4. Van Daele, G. H. P.; Bosmans, J.-P. R. M. A.; Schuurkes, J. A. J. PCT Int. Appl. **1996**, 15 pp., WO9616060

5. Sloots, C. E. J.; Poen, A. C.; Kerstens, R.; Stevens, M.; De Pauw, M.; Van Oene, J. C.; Meuwissen, S. G. M.; Felt-Bersma, R. J. F. *Aliment. Pharmacol. Ther.,* **2002**, *16*, 759-767
6. Krogh, K.; Jensen, M. Bach; Gandrup, P.; Laurberg, S.; Nilsson, J.; Kerstens, R.; De Pauw, M. *Scand. J. Gastroenterol.*, 2002, *37*, 431-436.
7. De Schryver, A. M. P.; Andriesse, G. I.; Samsom, M.; Smout, A. J. P. M.; Gooszen, H. G.; Akkermans, L. M. *Aliment. Pharmacol. Ther.,* **2002**, *16*, 603-612.
8. Bouras, Ernest P.; Camilleri, Michael; Burton, Duane D.; Thomforde, George; McKinzie, Sanna; Zinsmeister, Alan R. *Gastroenterology*, **2001**, *120*, 354-360.
9. Poen, A. C.; Felt-Bersma, R. J. F.; Van Dongen, P. A. M.; Meuwissen, S. G. M. *Aliment. Pharmacol. Ther.*, **1999**, *13*, 1493-1497.
10 Bouras, E. P.; Camilleri, M.; Burton, D. D.; McKinzie, S. *Gut*, **1999**,*44*, 682-686.
11. Marburg,S.; Tolman,R.L. *J. Het. Chem* **1980**, 1333-1335
12. Candian,I.; Debernardinis, S.; Cabri, W.; Marchi, M.; Bedeschi, A.; Penco, S.; *Synlett* **1993**, 269-270.
13. Sugasawa, T.; Toyoda, T.; Adachi, M.; Sasakura, K.; *J. Am. Chem. Soc.* **1978**, 4842-4852.
14. Plotkin, M.; Chen, S.; Spoors, P.G. *Tetrahedron Letters* **2000**, 41, 2269-2273.
15. Monte, A.P.; Marona-Lewicka, D.;Parker, M.A.; Wainscott, D.B.; Nelson, D.L.; Nichols, D.E. *J. Med. Chem.* **1996**,39,2953-2961.
16. Paquette, L.A.; Schulze, M.M.; Bolin, D.G. *J.Org. Chem.* **1994**,59,2043-2051.
17. Alabaster, R.J.; Cotrell, I.F.; Marley, H.; Wright, S.H.B. *Synthesis* **1988**, 950-952.
18. Bradsher, C.K.; Reames, D.C. *J.Org. Chem.* **1981**,46,1384-1388.
19. Knochel, P.; Jones, P.; *Organozinc reagents, a practical approach* **1999**, Oxford University Press.
20. Erdik, E.; *Tetrahedron* **1992**, 44, 9577-9648.
21. Rieke, R.D.; *Top. Curr. Chem.* **1975**, 59,1.
22. Gaudemar, M.; *Bull.Soc. Chim. Fr.***1962,** 974.
23. Erdik, E.; *Tetrahedron* **1987**, 43 (10), 2203-2212
24. Zhu, L.; Wehmeyer, R.M.; Rieke, R.D.; *J. Org. Chem.***1991**, 56, 1445-1453.

Chapter 9

Process Research Leading to an Enantioselective Synthesis of a 5-Lipoxygenase Inhibitor for Asthma

David B. Damon[1,*], Michael Butters[1,3], Robert W. Dugger[1], Peter Dunn[2], Sally Gut[1], Brian P. Jones[1], Thomas G. LaCour[1,4], C. William Murtiashaw[1,5], Harry A. Watson Jr.[1], James Weeks[1], and Timothy D. White[1]

[1]Chemical Research and Development, Pfizer Global Research and Development, Eastern Point Road, Groton, CT 06340
[2]Pfizer Global Research and Development, Building 180/2.19d, IPC 828, Sandwich, Kent, United Kingdom
[3]Current address: AstraZeneca Avlon Works, Severn Road, Hallen, Bristol BS10 7ZE, United Kingdom
[4]Current address: Department of Chemistry, University of California at Irvine, 4003 Reines Hall, Irvine, CA 92697
[5]Deceased October 25, 1995
***Corresponding author: telephone: 860–441–7923, fax: : 860–441–3630, email: david.b.damon@groton.pfizer.com**

(R)-*N*-{3-[3-(4-Fluorophenoxy)phenyl]-2-cyclopenten-1-yl}-*N*-hydroxyurea **1** is a 5-lipoxygenase inhibitor (5-LOI) recently under development for the treatment of asthma. Compound **1** is a member of the *N*-hydroxyurea class of 5-LOI's, and is distinguished by its central cyclopentene ring bearing a chiral, allylic *N*-hydroxyurea moiety and a pendant diarylether. This chapter discusses our process research on **1**, starting from the racemic discovery synthesis and culminating with an enantioselective route to **1** featuring early introduction of asymmetry, multiple crystalline intermediates, a high degree of convergency, and a minimum number of synthetic operations. The challenges encountered during our initial kilogram scale pilot plant campaign are presented, followed by a retrosynthetic analysis that led to the enantioselective approach to **1**. The realization of this approach is described in detail, highlighted by enantioselective 1,2-reduction of iodoenone **22**, enantioenrichment of Mitsunobu products **25** and **33** by recrystallization, and Suzuki coupling of **33** with boronic acid **19** to assemble the two halves of the molecule.

Introduction

(R)-*N*-{3-[3-(4-Fluorophenoxy)phenyl]-2-cyclopenten-1-yl}-*N*-hydroxyurea **1** is a 5-lipoxygenase inhibitor (5-LOI) recently under development for the treatment of asthma. 5-lipoxygenase (5-LO) is a key enzyme involved in the metabolism of arachidonic acid, a ubiquitous C-20 fatty acid found in a wide variety of mammalian cells.(*6*) 5-LO initiates an arachidonic acid metabolic cascade leading to the peptidoleukotrienes C4, D4 and E4 (LTC4, LTD4 and LTE4, respectively).(*6, 7*) The peptidoleukotrienes are potent mediators of bronchoconstriction and airway hypersensitivity reactions.(*7, 8*) Both 5-LOI's and LTD4 receptor antagonists have demonstrated clinical efficacy in the treatment of asthma.(*6, 9*) The 5-LOI zileuton and LTD4 antagonists pranlukast, zafirlukast and montelukast are currently marketed for this indication. Compound **1** is a member of the *N*-hydroxyurea class of 5-LOI's, (*10*) and is distinguished by its central cyclopentene ring bearing a chiral, allylic *N*-hydroxyurea moiety and a pendant diarylether. This chapter will discuss our process research on **1**, starting from the racemic discovery synthesis, going through the initial kilogram scale pilot plant campaign, and culminating with an efficient enantioselective synthesis of this molecule.

Figure 1

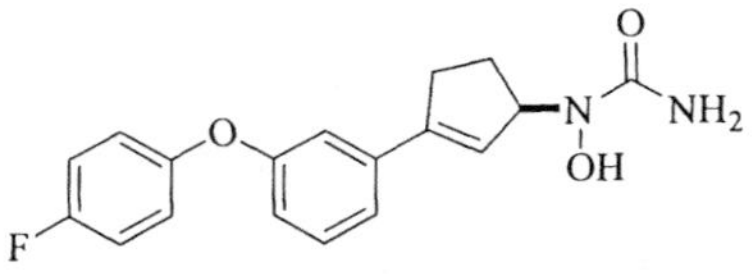

1

1. Discovery Synthesis

The discovery synthesis of **1** is outlined in Scheme 1. This route was sufficient to supply the needs of the discovery project team (pre-clinical pharmacologic, pharmacokinetic and drug safety studies) but required some modification prior to initial scale-up in our pilot plant. First, all intermediates in the synthesis (with the exception enone **4**) were oils requiring chromatographic purification. In the laboratory this was accomplished without difficulty, but on kilo scale chromatography can be a costly, labor intensive undertaking, particularly at the early stages of development where chromatography optimization studies have not been completed. Second, oxime **5** reduction used sodium cyanoborohydride.(*11*) This material, while perfectly adequate for use in the laboratory, requires extreme measures for use on large scale to prevent release of highly toxic hydrogen cyanide gas. In addition the yield was less than 60% and the product **6** required chromatographic purification. Finally and perhaps most significantly, an adequate resolution needed to be developed, which would allow for processing of multi-kilogram amounts of racemate.

Scheme 1

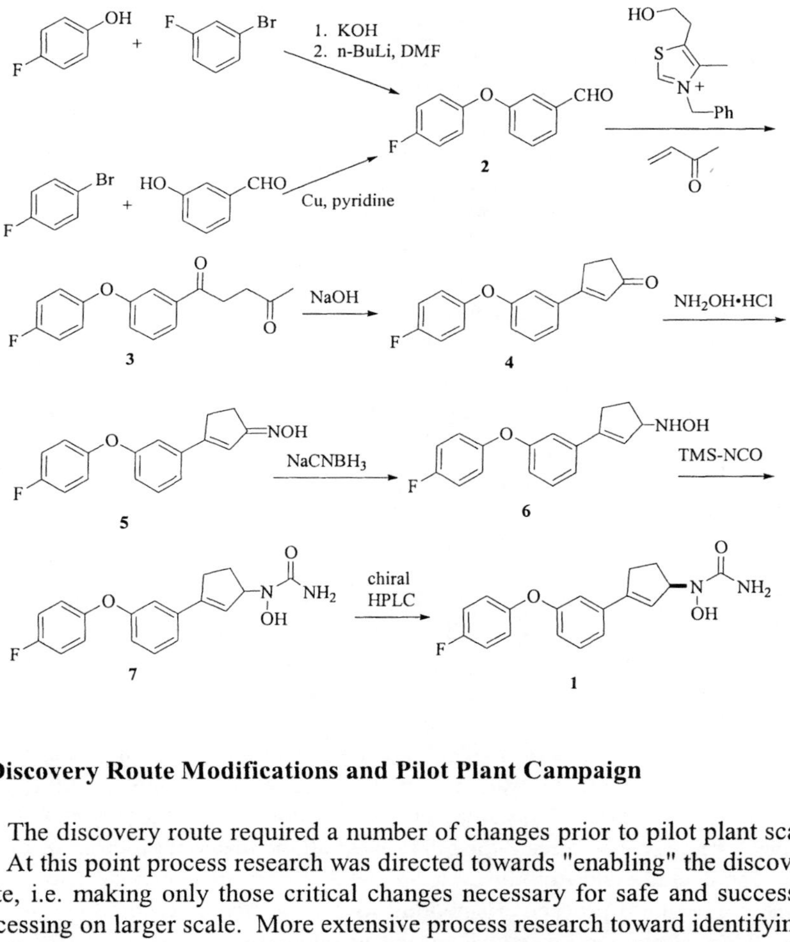

2. Discovery Route Modifications and Pilot Plant Campaign

The discovery route required a number of changes prior to pilot plant scale-up. At this point process research was directed towards "enabling" the discovery route, i.e. making only those critical changes necessary for safe and successful processing on larger scale. More extensive process research toward identifying a more efficient synthesis would wait until after completing the first scale-up. We kept the same bond disconnections as the discovery route for our first pilot plant campaign, setting our initial research goals to find a replacement for sodium cyanoborohydride and to find a scaleable resolution. The initial bulk synthesis is outlined in Scheme 2.

Scheme 2

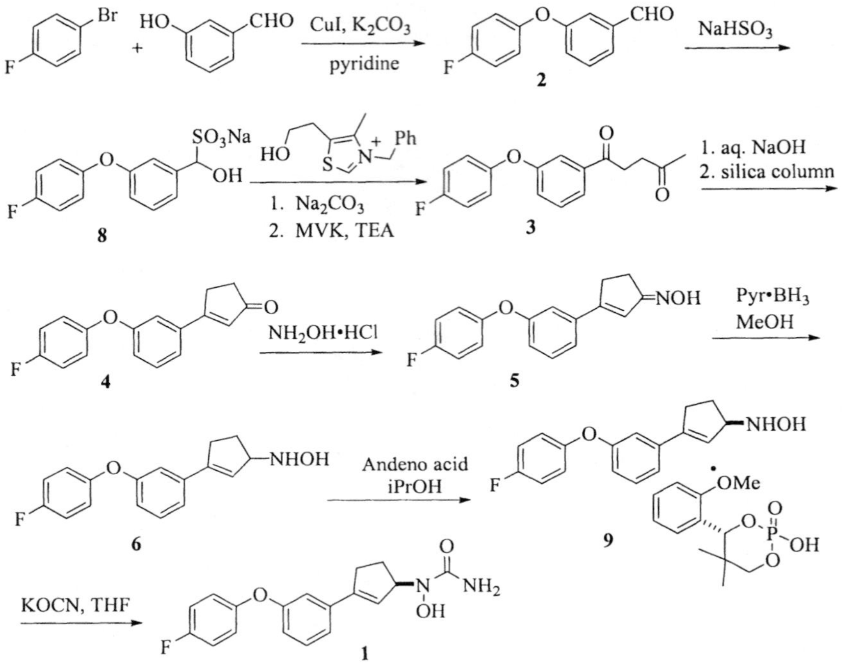

We used an Ullmann coupling (*12*) to prepare aldehyde **2** since this was one step less the alternate S_NAr/metallation approach (Scheme 1). The bisulfite adduct **8** was prepared to facilitate isolation (80% yield for the two steps). Breaking the bisulfite adduct liberated aldehyde **2**, which then underwent the Stetter (*13*) and aldol (*14*) reaction sequence as in the discovery synthesis. No purifications were done up to this point, and the purity of enone **4** was such that silica gel chromatography was required. The crude enone could not be crystallized, even after exhaustive attempts. This was a major bottleneck in this synthesis, requiring significant time to process all the material (approximately 50 kg!). The overall yield from bisulfite adduct **8** to the purified enone **4** was 31%. Oxime **5** formation proceeded smoothly to give nearly a quantitative yield of crystalline material. We encountered the second major bottleneck in the scale-up when we began the oxime reduction. In the laboratory we found conditions using borane-pyridine (*15*) in place of sodium cyanoborohydride to reduce oxime **5**. Although these new reduction conditions worked well in the laboratory, this reaction was extremely difficult to scale up. Yield and purity of product were capricious, varying considerably from batch to batch. The reaction had to be run at very low concentration. The starting oxime was not completely reduced under these conditions, typically stalling at about 75% conversion and thus requiring chromatographic purification. After considerable research, we determined that the principal cause for these problems was facile hydroxylamine **6** oxidation back to oxime **5**. We believe variability in degassing capabilities for different sized glassware and reactors accounted for the inconsistent results with these reactions. We were only able to successfully run the reduction up to 22 L scale and resorted to sixteen (!) 22 L runs to process all the material. Starting with 14.7 kg of oxime **5**, we isolated just 5.4 kg of hydroxylamine **6**. This one step was the major time and yield loss in the entire process.

A final processing bottleneck was resolution of hydroxylamine **6**, accomplished using the Andeno phosphoric acid.(*16*) This acid gave the best yield of resolved hydroxylamine after a thorough screen of resolving acids. Unfortunately this resolution was at best fair, both in terms of yield and enantiomeric excess. After salt formation and recrystallization we obtained only a 20% yield of hydroxylamine-Andeno acid salt **9** (of a maximum 50%) in only 80% ee. At the time we had no other resolution amenable to scale-up and could not further upgrade the enantiopurity (e.g. by further recrystallization of **9** or **1**). Having made it to the end of the synthesis, we were pleased when the last step proceeded smoothly and **1** was isolated after ethyl acetate recrystallization in good yield and chemical purity.

This route had several limitations precluding its use for further scale-up. First, multiple silica gel chromatographies were required due to the lack of sufficient crystalline intermediates. Second, the finicky oxime reduction could not be scaled beyond 22 L. Third, the resolution occurred late in the synthesis, required an expensive resolving agent (at the time ~$6,000/kg), and proceeded in low yield and ee.

3. Early Process Research

As discussed in section 2 above, the oxime reduction was the major bottleneck in our pilot plant campaign. At one point while the campaign was underway we were not confident the oxime reduction could be accomplished on large scale. Consequently we researched alternate routes to hydroxylamine **6** which did not require this troublesome reaction. All new approaches used allylic alcohol **10** as a precursor to hydroxylamine **6**. Compound **10** could be prepared in high yield by Luche reduction (*17*) of enone **4**. We examined three different ways to convert the allylic alcohol **10** to hydroxylamine **6**: first, a Mitsunobu reaction (*18*) based approach using *N*-BOC-*O*-BOC-hydroxylamine as nucleophile followed by subsequent BOC deprotection; second, a palladium catalyzed process to install *N*-BOC-*O*-BOC-hydroxylamine, (*19*) followed by deprotection; third, a solvolytic approach to displace the hydroxy group with an *O*-protected hydroxylamine nucleophile (*20*) followed by deprotection.

Scheme 3

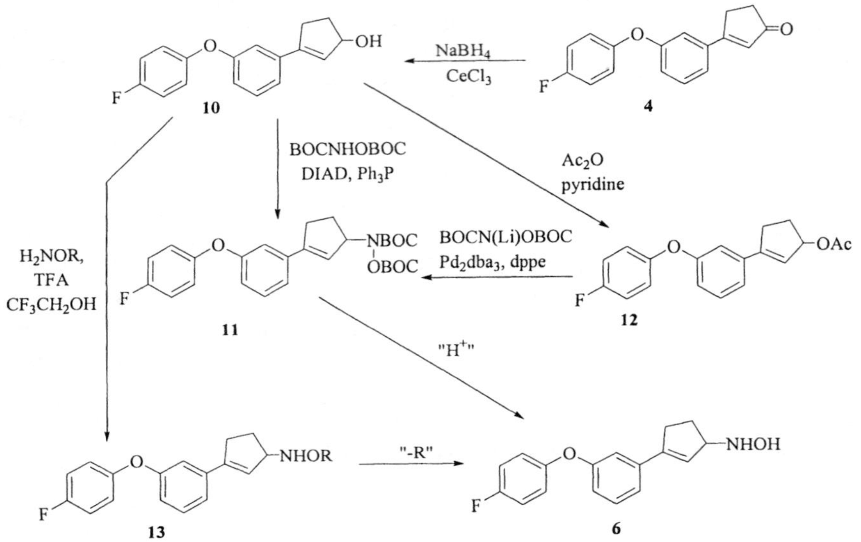

All three approaches had the advantage of replacing the oxime reduction with a straightforward ketone reduction. The Mitsunobu reaction proceeded in moderate yield. Product **11** was an oil requiring chromatographic purification. The palladium catalyzed reaction with **12** similarly went in moderate yield and required chromatography. We hoped to eventually use an enantioselective version of the palladium chemistry [à la Trost (*21*)], but abandoned this idea because we were unable to find clean BOC deprotection conditions. The optimal conditions identified used trimethylsilyl triflate and 2,6-lutidine in methylene chloride and gave at best 70% yield of **6** after chromatography. All other acidic conditions resulted in decomposition of **6**, underscoring the acid lability of this allylic substituted cyclopentene system.

The solvolytic approach was intriguing due to its relative simplicity compared with either the Mitsunobu or palladium catalyzed processes. *O*-alkyl protected hydroxylamines [for example *O*-benzylhydroxylamine and *O*-4-methoxybenzylhydroxylamine (*22*)] could add in good yield and high regioselectivity via this method.(*23*) Successful deprotections were not achieved with the *O*-alkyl protected substrates. *O*-silyl protected hydroxylamines also added in high regioselectivity, but yields were compromised by concomitant silyl deprotection and oxime formation *in situ*. The best protecting group found, *O-t*-butyldiphenylsilylhydroxylamine (*24*) gave about 75% overall yield of **6** contaminated with 10-15% of oxime **5** after *in situ* deprotection.

These three alternate approaches to hydroxylamine **6** were dropped for two reasons: first, they offered limited advantage over existing methods; second, improved oxime reduction conditions were found which enabled processing of the batch along the original synthetic route (albeit only on 22 L scale).

4. The First Enantioselective Route

Clearly the existing route could not be scaled up beyond the current level. This was highlighted during the planning for a proposed 10 kg campaign, which would among other drawbacks require purchase of over one metric ton of silica gel! Upon completion of the bulk campaign we looked for a better synthesis of **1**. Our goals were to find a convergent route featuring crystalline intermediates, no chromatography and early introduction of asymmetry. Our retrosynthesis is outlined in Scheme 4.

Scheme 4

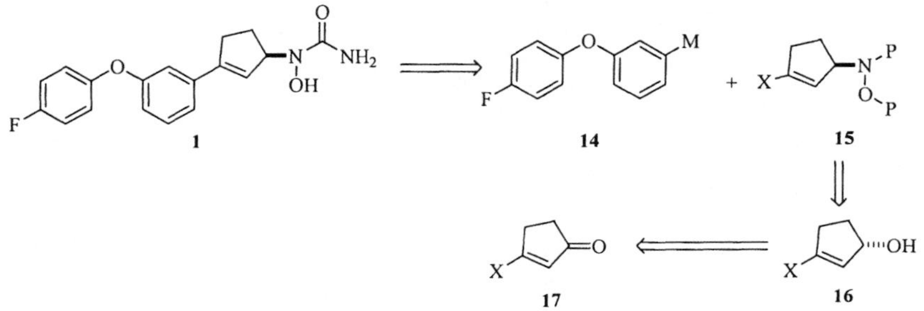

In a retrosynthetic sense, breaking the cyclopentene-aryl bond seemed very attractive, forming two roughly equal halves of the molecule. Compounds **14** and **15** could be joined synthetically using a cross coupling reaction where X is a halide or triflate and M is a metal such as boron or zinc. Protected hydroxylamino-cyclopentene **15** could arise from S_N2 inversion of chiral cyclopentenol **16**, which in turn could be formed by chiral reduction (or achiral reduction/resolution) of 3-substituted cyclopentenone **17**. The development of a new route based on this retrosynthetic analysis is described below.

We elected to use Suzuki reaction (*25*) to couple the two halves of the molecule. From a process point of view this method offered several advantages over other cross coupling reactions [e.g. Stille, (*26*) Negishi, (*27*) or Kumada (*28*)]: a stable organometallic intermediate can be prepared and isolated (boronic acid), mild reaction conditions are used (no cryogenic temperature), and heavy metals (e.g. tin) are avoided. The synthesis of the aryl ether boronic acid **19** (*29*) proceeded in straightforward fashion, and is depicted in Scheme 5.

Scheme 5

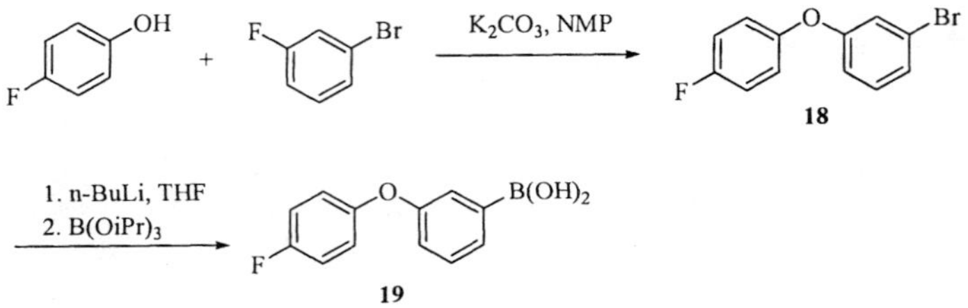

OH F + F Br K_2CO_3, NMP → F O Br **18**

1. n-BuLi, THF 2. $B(OiPr)_3$ → F O $B(OH)_2$ **19**

We chose 3-iodo-cyclopent-2-enone **22** as the starting point for the other half of the molecule.(*30*) This is a known compound, prepared by treatment of 1,3-cyclopentanedione with triphenylphosphine and iodine.(*31*) We developed a complementary two-step, one pot process to synthesize **22** by *in situ* activation with mesyl chloride followed by addition of iodide and elimination.(*32*)

Scheme 6

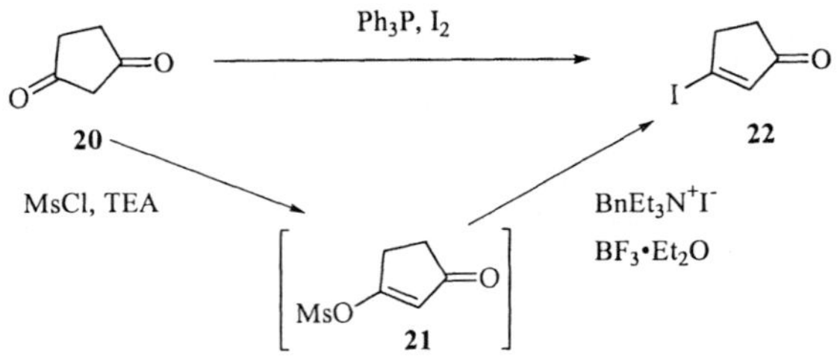

O =O **20** → Ph_3P, I_2 → I =O **22**

MsCl, TEA → [MsO =O **21**] → $BnEt_3N^+I^-$ $BF_3{\bullet}Et_2O$

Next came the challenging task of introducing the hydroxylamine moiety in an asymmetric fashion. To accomplish this we first needed access to enantiomerically enriched/pure 3-iodocyclopent-2-enol (**24**), then we needed to establish that this allylic alcohol would undergo S_N2 reaction with a protected hydroxylamine nucleophile without loss of stereochemical integrity. We found that racemic 3-iodo-cyclopent-3-enol **23** could be prepared in high yield by Luche reduction (*17*) of enone **22**. After screening a number of lipases we found that Amano Lipase P selectively acylated the undesired alcohol enantiomer, affording the desired (S) alcohol **24** in 40% yield and >99 % ee.(*33*)

Although the lipase resolution results looked promising, direct enantioselective 1,2-reduction of the enone would be much more efficient if appropriate conditions could be found. It was known that Corey's oxazaborolidine (CBS) catalyst would reduce 2-bromo-cyclopent-2-enone with high enantioselectivity (*34*) due to the large bromine atom being located alpha to the ketone. It was uncertain if **22** would reduce with enantioselectivity since the iodine atom was one atom further removed from the ketone. After an extensive effort at this reduction, our best conditions gave a 90% yield of enantioenriched **24** (3:1 S:R) using 5 mol % (*R*) *B*-methyl CBS catalyst.(*35*)

Scheme 7

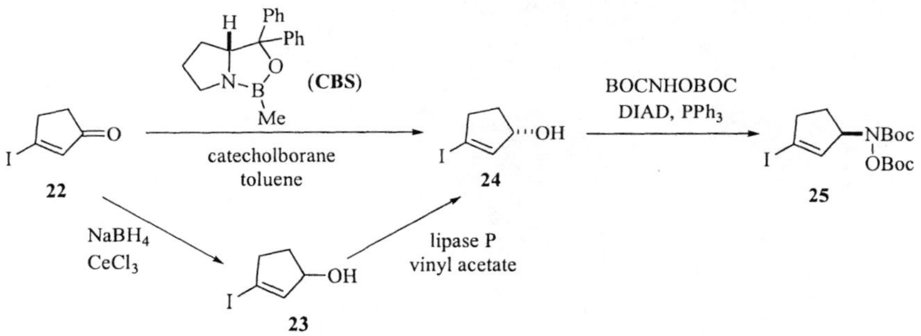

H Ph Ph O N B Me (CBS)
catecholborane
toluene
BOCNHOBOC
DIAD, PPh_3
I O **22**
I OH **24**
I NBoc OBoc **25**
$NaBH_4$
$CeCl_3$
lipase P
vinyl acetate
I OH **23**

Our initial plan was to use the CBS reduction to generate **24** with a 3:1 ratio of enantiomers, then use the lipase kinetic resolution to remove the minor undesired enantiomer. This could be accomplished conveniently in a one-pot protocol, as both the CBS reduction and the lipase resolution were run in toluene. Before we did this, however, we wanted to assure ourselves that a protected hydroxylamine nucleophile could add to the enantioenriched allylic alcohol **24** without eroding ee.A preliminary coupling reaction used a sample of **24** partially resolved from a lipase resolution (84% ee). This material was subjected to Mitsunobu conditions (*36*) with *N*-BOC-*O*-BOC-hydroxylamine to give about an 80 % yield of product **25**, shown to have 80% ee by chiral HPLC analysis. We were encouraged that very little erosion of stereochemical integrity was observed under the Mitsunobu introduction of the protected hydroxylamine nucleophile. A sample of the 80% ee **25** was then recrystallized for single crystal x-ray analysis (at the time we needed absolute stereochemical information on our new intermediates). To our great surprise, two different crystals grew upon recrystallization: large, ~1 mm diameter rods and small, thin feathery needles (see Figure 2 for a photomicrograph of these crystals). After mechanical

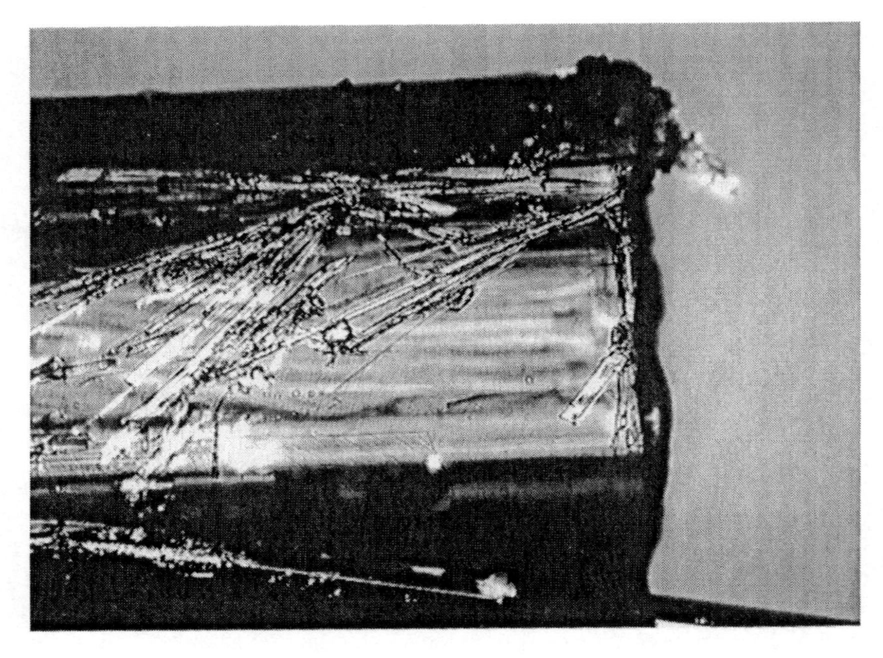

***Figure 2:** photomicrograph of compound **25** crystals*

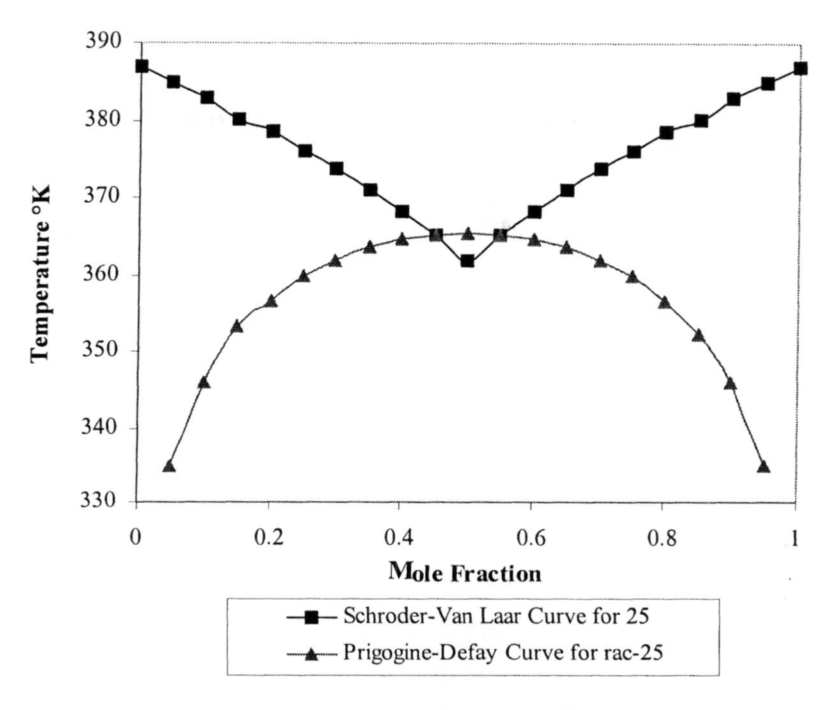

Figure 3: *Binary Phase Diagram for **25** and rac-**25***

separation of the crystals, chiral HPLC showed the large rods to be enantiomerically pure, while the thin needles were racemic. Single crystal x-ray confirmed the large rods to be the desired enantiomer.

This result demonstrated that 80% ee **25** could be upgraded to higher enantiopurity by recrystallization. Next we wanted to learn how low the ee level in **25** could be and still obtain enantioenrichment by recrystallization. To do this we prepared a binary phase diagram (*37*) for **25** and *rac*-**25**, shown in Figure 3.

This diagram indicates that **25** at greater than 10% ee should show enantioenrichment upon recrystallization. Using doped samples we showed that while this is true, at least 40% ee is needed to get practical yields of **25** with high enantiopurity. Since CBS reduction of **22** gave 50% ee **24** and Mitsunobu conversion of **24** to **25** maintained this purity level, we would not need a lipase process to produce enantiopure **25**. We could achieve this simply by recrystallization. This was reduced to practice by carrying crude product from the CBS reduction into the Mitsunobu reaction, which after a reslurry (to remove triphenyl phosphine oxide and reduced diisopropylazodicarboxylate) and two crystallizations afforded **25** in 40% yield and >96% ee.

Scheme 8

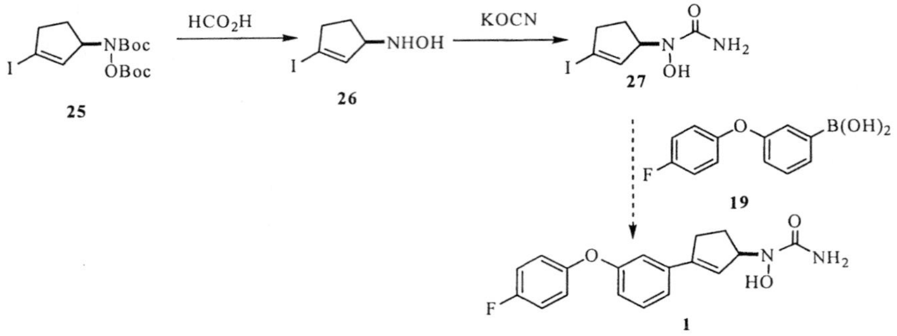

Our initial end-game plan was to convert Mitsunobu product **25** to *N*-hydroxyurea **27**, then couple **27** with boronic acid **19** to generate **1** (Scheme 8).

Coupling **27** and **19** could be accomplished on small lab pilots, however never in greater than 20% yield. We attributed this to the *N*-hydroxyurea moiety chelating palladium and shutting down the catalytic cycle. Coupling Mitsunobu product **25** with boronic acid **19** was not an option, since we knew from our earlier work (section 3) that the two BOC groups would be very difficult to remove cleanly. After quite a bit of research we found that protecting the *N*-hydroxyurea oxygen enabled a successful Suzuki reaction to occur.(*38*) We needed a protecting group stable to the Suzuki conditions and removable at the end of the synthesis. Compound **1** is unstable to acidic and hydrogenation conditions (due to elimination and olefin reduction, respectively), limiting us to neutral or basic conditions for the final deprotection. We found that an *O*-BOC group was surprisingly unstable to the Suzuki conditions when run at elevated temperatures (70°C). In fact, the *O*-BOC group of coupled product **29** could be removed by base treatment (e.g. K_2CO_3 and methanol), no doubt facilitated by the excellent leaving group properties of the *N*-hydroxyurea moiety. Running the Suzuki reaction at room temperature prevented *O*-BOC cleavage and gave essentially a quantitative yield of **29**.

Scheme 9

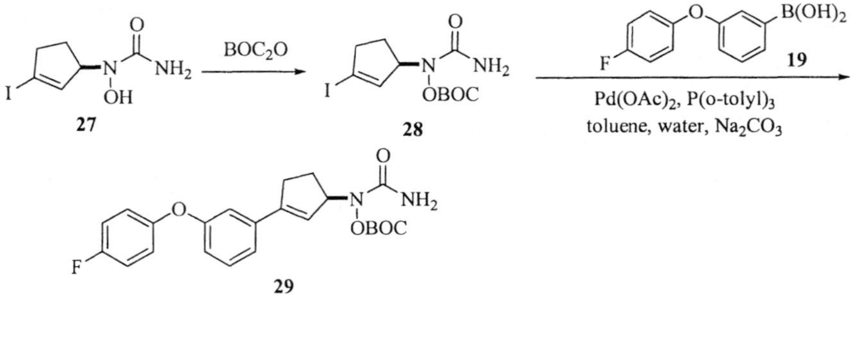

All that remained was to optimize conditions to remove the *O*-BOC group from **29** to generate **1**. We found that **29** underwent two reaction pathways when treated with base. The major one resulted from desired nucleophilic attack at the BOC carbonyl leading to **1**. The minor one arose from basic attack at the urea NH_2, leading to BOC migration from oxygen to nitrogen followed by *in situ* oxadiazodione formation. Once formed, the oxadiazodione was resistant to base hydrolysis. K_2CO_3 and methanol, for example, gave a 2:1 ratio of **1**:**31**.

Scheme 10

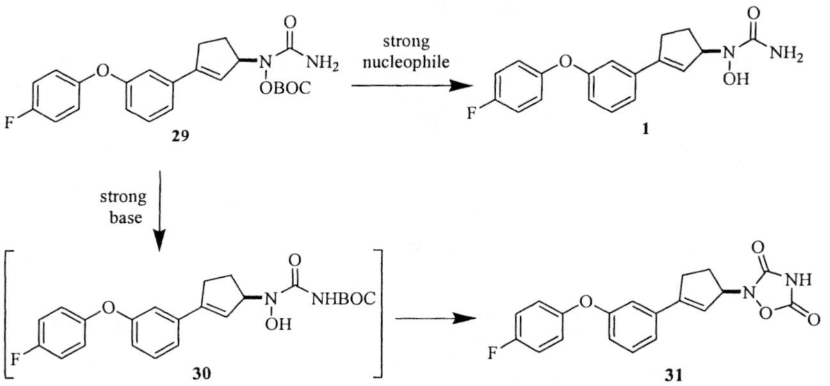

We felt that using a nucleophile stronger than methoxide would favor the desired reaction pathway. After a screen of various bases and solvents, (*39*) we found that hydroxylamine and hydrazine favored the nucleophilic pathway with minimal oxadiazodione formation. For worker safety reasons we chose to use hydroxylamine for this transformation.(*40*) Our optimized conditions treated **29** with excess hydroxylamine hydrochloride and TEA in ethanol at room temperature and gave an 85% recrystallized yield of **1**.

Scheme 11

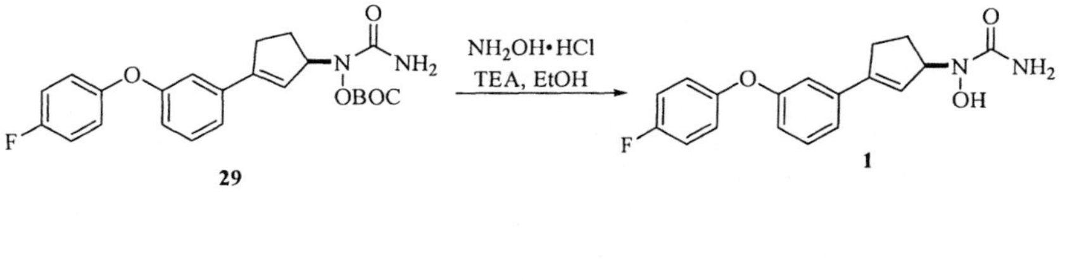

The first enantioselective route to **1** offered several advantages over the existing route: many more crystalline intermediates, early introduction of asymmetry, no oxime reduction, convergency, and no chromatography. There were still some problems that needed fixing, in particular the extra BOC group manipulations and only moderate selectivity on the CBS reduction. Section 6 describes the process research that addressed these issues, culminating in the second enantioselective synthesis of **1**. Section 5 describes a key result that helped us get there.

5. A Second Racemic Route, and a Key Result

The project required additional bulk material at a date earlier than the first enantioselective route could be optimized for scale-up. The goal was to identify a route capable of delivering bulk material in a minimal amount of time. To support this short-term need we developed a second racemic route to **1**, outlined in Scheme 12.

Scheme 12

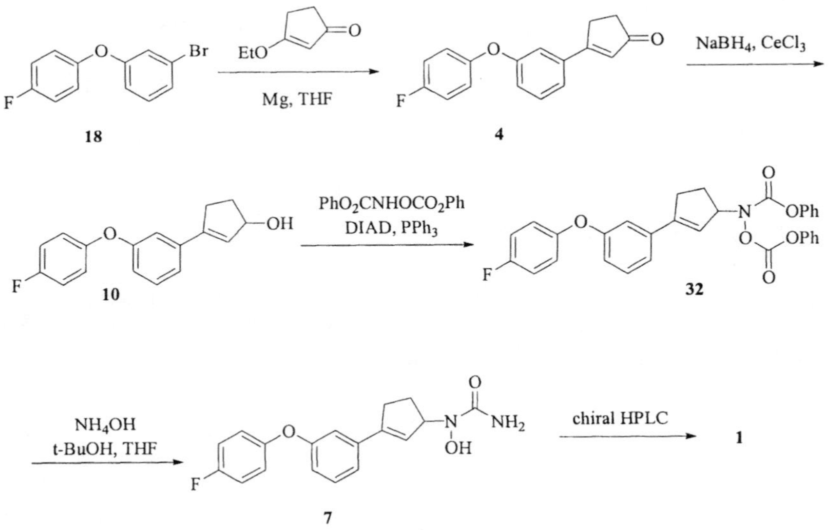

This route began with formation of the aryl Grignard reagent from **18** followed by addition of 3-ethoxy-cyclopent-2-enone (*41*) to form enone **4** in over 80% yield.(*42*) A Luche reduction (*17*) formed allylic alcohol **10**, which was subjected to Mitsunobu conditions with *N,O*-

Scheme 13

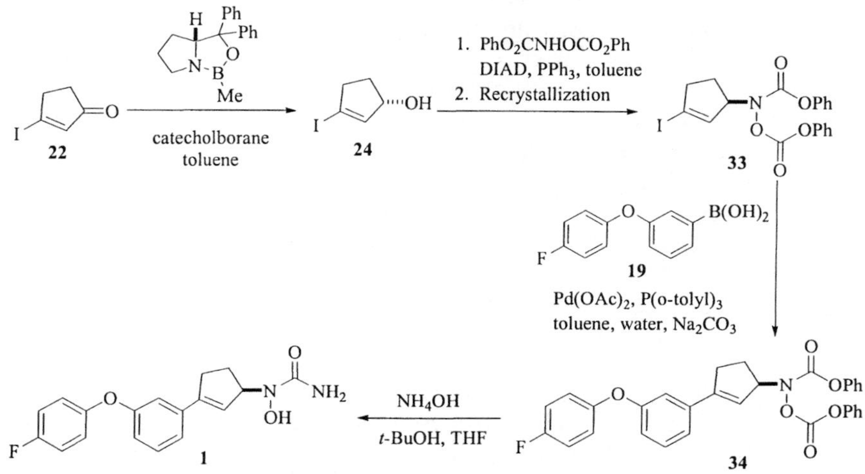

bisphenoxycarbonylhydroxylamine (*43*) as nucleophile. The Mitsunobu product **32** was treated with ammonia (*44*) to afford **7** in about 30% overall yield from **4**.(*45*) By this point in the project we had developed an efficient chromatographic resolution of **1** which we planned to use for immediate bulk needs. This process was successfully demonstrated on 22 L scale. This route was very short and, despite the low yielding Mitsunobu reaction and last step resolution, allowed for efficient processing of **1**. More importantly, this route demonstrated the utility of *N,O*-bisphenoxycarbonylhydroxylamine as an *N*-hydroxyurea synthon in the preparation of **1**. This key result led to the second enantioselective route, described in section 6.

6. The Second Enantioselective Route

From a synthetic efficiency point of view the first enantioselective route had two limitations, first the moderate enantioselectivity obtained from CBS reduction of iodoenone **22** and second the extra steps spent manipulating the *O*-BOC group. We reinvestigated the chiral reduction of iodoenone **22** with the goal of improved enantioselectivity. We envisioned that replacement of *N*-BOC-*O*-BOC-hydroxylamine with *N,O*-bisphenoxycarbonylhydroxylamine in the Mitsunobu step could eliminate the extra deprotection/protection steps. The major question was whether Mitsunobu product **33** would show ee enrichment as

25 had upon recrystallization. These goals were realized, as depicted in the second enantioselective route shown below.

We first examined other chiral reductions of iodoenone **22** with the goal of achieving greater than 50% ee. The Quallich catalyst (*46*) gave no enantioselectivity at room temperature and no reaction at low temperatures. Darvon alcohol modified LAH reagent (*47*) gave a slight excess of the undesired enantiomer at low temperatures and decomposed upon warming. More promising was a diaminoalcohol modified LAH reagent (*48*) that gave **24** with 70% ee in over 80% yield. We also reexamined the CBS reduction itself, and found that running at higher catalyst load (20% versus 5%) and lower temperatures (-70°C throughout) gave **24** with 70% ee in 87% yield. We chose to stay with the CBS catalyst since it is available commercially, while the diaminoalcohol ligand and LAH complex requires preparation.

The Mitsunobu reaction converting **24** to **33** was carried out using identical conditions as before, but substituting *N*,*O*-bisphenoxycarbonyl-hydroxylamine for *N*-BOC-*O*-BOC-hydroxylamine. Chemical yields were comparable and **34** showed no loss in enantiopurity. Compound **34** could be enriched to >96% ee by single recrystallization from ethanol-water. Purified **34** then underwent Suzuki coupling with boronic acid **19** as described in section 4, and was converted to **1** in high yield by ammonia treatment as described in section 5. At this point we reached our goal of defining an enantioselective route to **1** featuring early introduction of asymmetry, multiple crystalline intermediates, a high degree of convergency (the longest linear sequence is 5 steps), and a minimum number of synthetic operations.

Conclusions

This chapter described process research leading to an enantioselective synthesis of **1**. The road to developing new chemical processes contains many twists and turns, detours, and short cuts that often end up becoming long cuts. Such was the case here. Each stage of the journey (from enabling the discovery route and completing the kilo scale campaign, to developing a new racemic route and two new enantioselective routes) was accomplished thanks to close collaboration between our research and scale-up groups, empiricism, hard work and a little serendipity. This process research story is an example of the fundamental role synthetic organic chemistry plays in the development of new drug candidates.

References and Notes

1. Author to whom correspondence should be addressed: Pfizer Global Research and Development, MS 8156-117, Groton, CT 06340, USA. Phone: (860) 441-7923, FAX: (860) 441-3630, e-mail: david_b_damon@groton.pfizer.com
2. Current address: AstraZeneca Avlon Works, Severn Road, Hallen, Bristol BS10 7ZE United Kingdom.
3. Pfizer Global Research and Development, Bld. 180/2.19d, IPC 828, Sandwich, Kent, United Kingdom.
4. Current address: Department of Chemistry, University of California, Irvine, 4003 Reines Hall, Irvine CA 92697
5. Deceased October 25, 1995
6. O'Byrne, P.M.; Israel, E.; Drazen, J.M. *Ann. Intern. Med.* **1997**, *127*, 472-480.
7. Steinhilber, D. *Current Med. Chem.* **1999**, *6*, 71-85.
8. Adams, J.L.; Garigipati, R.S.; Sorenson, M.; Schmidt, S.J.; Brian, W.R.; Newton, J.F.; Tyrrell, K.A.; Garver, E.; Yodis, L.A.; Chabot-Fletcher, M.; Tzimas, M.; Webb, E.F.; Breton, J.J.; Griswold, D.E. *J. Med. Chem.* **1996**, *39*, 5035-5046 and references therein.
9. Chambers, R.J.; Marfat, A. *Exper. Opin. Ther. Patents*, **1999**, *9*(1), 19-26.
10. (a) Bell, R.L. In *Novel Inhibitors of Leukotrienes*; Folco, G.; Samuelsson, B.; Murphy, R.C. Eds.; Birkhaeuser Verlag: Basel, Switzerland, 1999; pp 235-249. publisher. (b) Thomas, A.V.; Patel. H.H.; Reif, L.A.; Chemburkar, S.R.; Sawick, D.P.; Shelat, B.; Balmer, M.K.; Patel, R.R. *Org. Proc. Res. and Dev.* **1997**, *1*, 294-299. (c) Flisak, J.R.; Lantos, I.; Liu, L.; Matsuoka, R.T.; Mendelson, W.L.; Tucker, L.M.; Villani, A.J.; Zhang, W.-Y. *Tetrahedron Lett.* **1996**, *37*(27), 4639-4642.
11. Borch, R.F.; Bernstein, M.D.; Durst, H.D. *J. Am. Chem. Soc.* **1971**, *93*(12), 2897-2904.
12. Theil, F. *Angew. Chem. Int. Ed.* **1999**, *38*(16), 2345-2347.
13. Stetter, H. *Angew. Chem. Intl. Ed.* **1976**, *15*(11), 639-647.
14. Novák, L.; Rohály, J.; Gálik, G.; Fekete, J.; Varjas, L.; Szántay, C. *Liebigs Ann. Chem.* **1986**, *3*, 509-524.
15. (a) Andrews, G.C. In *Encyclopedia of Reagents for Organic Synthesis*; Paquette, L. Ed.; Wiley: Chichester, 1995; Vol. 1, pp 637-638. (b) Kikugawa, Y.; Kawase, M. *Chemistry Lett.* **1977**, 1279-1280.
16. (a) Ten Hoeve, W.; Wynberg, H. *J. Org. Chem.* **1985**, *50*(23), 4508-14. (b Hulshof, L.A.; Broxterman, Q.B.; Vries, T.R.; Wijnberg, H.; Van Echten, E. EP 838448, 1998. (c) Both enantiomers are commercially available fromAldrich.
17. Luche, J.-L.; Gemal, A.L. *J. Am. Chem. Soc.* **1981**, *103*, 5454-5459.

18. (a) Mitsunobu, O. *Synthesis*, **1981**, 1-28. (b) Hughes, D.L. *Org. Prep and Proc. Intl.* **1996**, *28*(2), 127-164. (c) Malamas, M.S.; Palka, C.L. *J. Het. Chem.* **1996**, *33*(2), 475-478.
19. (a) Genet, J.-P.; Thorimberi, S.; Touzin, A.M. *Tetrahedron Lett.* **1993**, *34*(7), 1159-1162. (b) Connell, R.D.; Rein, T.; Aakermark, B.; Helquist, P. *J. Org. Chem.* **1988**, *53*(16), 3845-3849.
20. Adams, J.L.; Garigipati, R.S.; Griswold, D.E.; Schmidt, S.J. WO 91 14674, 1991
21. Trost, B.M.; Van Vranken, D.L. *Chem. Rev.* **1996**, *96*, 395-422.
22. Martin, S.F.; Oalmann, C.J.; Liras, S. *Tetrahedron*, **1993**, *49*(17), 3521-3532.
23. No regioisomer was detected by ^{1}H NMR or HPLC.
24. Stewart, A.O.; Martin, J.G. *J. Org. Chem.* **1989**, *54*(5), 1221-1223.
25. Miyaura, N.; Suzuki, A. *Chem. Rev.* **1995**, *95*, 2457-2483.
26. Farina, V.; Krishnamurthy, V.; Scott, W.J. In *Organic Reactions*; Paquette, L. Ed.; Wiley: New York, 1977; Vol. 50, pp 1-652.
27. Stanforth, S.P. *Tetrahedron*, **1998**, *54*(3/4), 263-303.
28. Kumada, M. *Pure and Appl. Chem.* **1980**, *52*(3), 669-679.
29. Webber, S.E.; Canan-Koch, S.S.; Tikhe, J.; Thoresen, L.H. WO 00 42040, 2000.
30. We prepared the corresponding bromides and found them to be much less stable than the iodo intermediates.
31. Piers, E.; Grierson, J.R.; Lau, C.K.; Nagakura, I. *Can. J. Chem.* **1982**, *60*(2), 210-223.
32. Kowalski, C.J.; Fields, K.W. *J. Org. Chem.* **1981**, *46*(1), 197-201.
33. Reaction conditions: **23**, toluene (10 mL/g of **23**), Amano lipase P (25 wt %), vinyl acetate (1.0 equiv), room temperature.
34. (a) Corey, E.J.; Chen, C.-P.; Reichard, G.A. *Tetrahedron Lett.* **1989**, *30*(41), 5547-5550. (b) Corey, E.J.; Rao, K.S. *Tetrahedron Lett.* **1991**, *32*(36), 4623-4626. (c) Corey, E.J.; Helal, C.J. *Angew. Chem. Int. Ed.* **1998**, *37*, 1986-2012.
35. Reaction conditions: **22**, toluene (20 mL/g of **22**), (*R*)-B-methyl CBS catalyst (0.05 equiv), catechol borane (1.1 equivalent), -70°C to -30°C.
36. Reaction conditions: **24**, toluene (20 mL/g of **24**), diisopropylazodicarboxylate (1.0 equivalent), triphenylphosphine (1.0 equivalent), N-BOC-O-BOC-hydroxylamine (1.0 equivalent), -70°C.
37. (a) Jacques, J.; Collet, A.; Wilen, S.H. *Enantiomers, Racemates and Resolutions*; Wiley: New York, 1981; Chapter 2. (b) Ḱozma, D.; Pokol, G.; Ács, M. *J. Chem. Soc. Perkin Trans. 2*, **1992**, *3*, 435-439. (c) Wilen, S.H.; Collet, A.; Jacques, J. *Tetrahedron* **1977**, *33*, 2725-2736.
38. The O-methyl ether of **27** was prepared as a model substrate to determine which part of the N-hydroxy urea required protection. High yield Suzuki

couplings were achieved with this compound, prompting our search for a suitable oxygen protecting group.

39. Alkali metal carbonates in methanol gave 2-4:1 ratios of **1**:**31**. Alkali metal hydroxides (aqueous in THF or anhydrous in DMSO) gave almost exclusively **31**. Amines (TEA, ammonia or DMAP) in methanol favored **1** but gave poor conversion.
40. Hydrazine is a suspected carcinogen. Inhalation of its vapors or ingestion of the liquid can cause nausea, vomiting, dizziness and convulsions. see Patnaik, P. *Hazardous Properties of Chemical Substances*; Wiley: New York, 1999; pp 827-828.
41. Banwell, M.G.; Harvey, J.E.; Jolliffe, K.A. *J. Chem. Soc., Perkin Trans. 1*, **2001**, *17*, 2002-2005.
42. No significant amount of 1,2-addition occurred, as evidenced by analysis of HPLC data.
43. Stewart, A.O.; Brooks, Dee. W. *J. Org. Chem.* **1992**, *57*(18), 5020-5023.
44. This is a known protocol to prepare *N*-hydroxyureas. See reference 42.
45. The Mitsunobu reaction was responsible for the relatively low yield. The Luche reduction and ammonia deprotection reactions both proceeded in high yield.
46. Quallich, G.J.; Woodall, T.M. *Synlett*, **1993**, 929-930.
47. Cohen, N.; Lopresti, R.J.; Neukom, C.; Saucy, G. *J. Org. Chem.* **1980**, *45*(4), 582-588.
48. (a) Sato, T.; Gotoh, Y.; Wakabayashi, Y.; Fujisawa, T. *Tetrahedron Lett.* **1983**, *24*(38), 4123-4126. (b) Sato, T.; Goto, Y.; Fujisawa, T. *Tetrahedron Lett.* **1982**, *23*(40), 4111-4112.

Chapter 10

A Practical Synthesis of Tetrasubstituted Imidazole p38 MAP Kinase Inhibitors: A New Method for the Synthesis of α-Amidoketones

Jerry A. Murry, Doug Frantz, Lisa Frey, Arash Soheili, Karen Marcantonio, Richard Tillyer, Edward J. J. Grabowski, and Paul J. Reider

Process Research Department, Merck Research Laboratories, Merck and Company, Inc., P.O. Box 2000, Rahway, NJ 07065

In this manuscript we disclose new synthetic methodology to prepare a member of a class of tetrasubstituted imidazole p38 inhibitors. The optimal route involves a thiazolium catalyzed cross acyloin-type condensation of a pyridinealdehyde with an *N*-acylimine. The pyridinealdehyde was prepared in 3 steps and 68% yield from 2-chloro-4-cyano pyridine. The tosylamide precursor to the *N*-acyl imine was prepared in two steps and 93% yield from isonipecotic acid. We have demonstrated the scope and some preliminary mechanistic studies concerning this new reaction. The resulting α-keto-amide is then cyclized with methyl ammonium acetate to provide the desired tetrasubstituted imidazole. Cbz deprotection and formation of a pharmaceutically acceptable salt completes the synthesis in 6 steps and 38% overall yield.

Introduction

The pyridinylimidazoles are representative of a class of polysubstituted imidazoles which are potent inhibitors of p38 Mitogen-Activated Protein (MAP) kinase.(*1*, 2) (Figure 1) These compounds have been indicated for the treatment of inflammatory diseases due to their mode of action which involves upstream regulation of several pro-inflammatory cytokines such as IL-1 (interleukin-1) and TNF (tumor necrosis factor).(*3*) Effective syntheses for this class of compounds have been described.(*4, 5*)

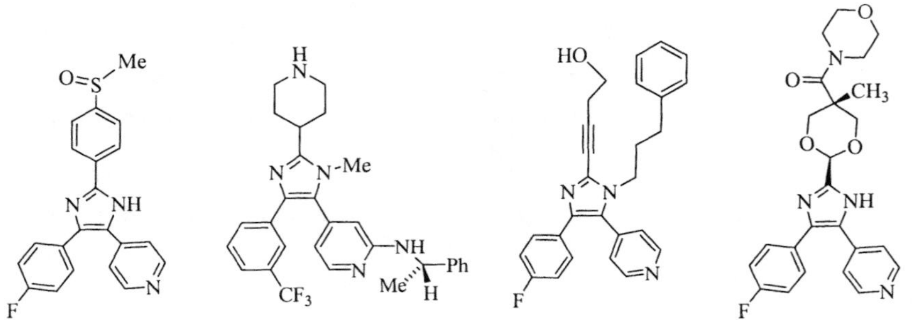

SB-203,580 1 RWJ-67657 RPR-200765A

***Figure 1**. Pyridylimidazole p-38 Kinase inhibitors*

In this chapter we report new methodology which allows for an efficient, practical synthesis of the p38 MAP kinase inhibitor **1**.(*6, 7*) The key step in this synthesis is a thiazolium catalyzed cross-acyloin coupling reaction of an aldehyde with an acyl imine.(*8*)

The previously reported syntheses of these compounds have relied on preparing a trisubstituted imidazole and then adding the fourth substituent either by *N*-alkylation or metallation of the 2-position and addition of an electrophile. The *N*-alkylations may suffer from lack of regioselectivity that usually results in loss of yield and the need for chromatographic separations. However, functionalization of the 2-position has been reported to proceed in high yields. We initiated a program aimed towards an efficient synthesis of this class of compounds that would provide tetrasubstituted imidazoles rapidly and be amenable to large-scale synthesis.

Route Selection

We initially investigated numerous routes towards preparing poly-substituted imidazoles. Four of those routes will be discussed in this section.

Tosylmethyl Isocyanide (TOSMIC) Dipolar Cycloaddition

We investigated an approach towards the synthesis of **1** utilizing the tosylmethyl isocyanide (TOSMIC) methodology and the results are outlined in Scheme 1.(*9*)

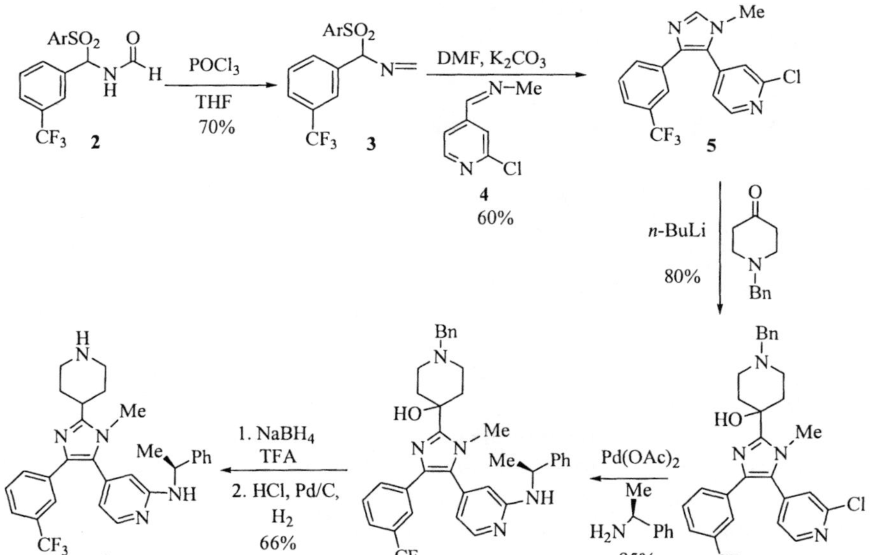

Scheme 1*. TOSMIC approach to* ***1***

Preparation of the requisite Tosyl formamide **2** using the procedure of Sisko (*4, 5*) occurred in good yield. Dehydration of **2** with $POCl_3$ provided the TOSMIC reagent **3** which underwent 1,3 dipolar cycloaddition with methyl imine **4** to provide imidazole **5**. However, this procedure proved problematic on larger scale. The reactions turned very dark and formed several unidentified by-products. Fortunately, the resulting mixture could be purified by silica gel chromatography to provide the desired trisubstituted imidazole in 60% yield. The installation of the piperidine substituent in the 2-position was effected in a two-step sequence. Lithiation with *n*-BuLi followed by addition of *N*-benzyl piperidone provided the desired tertiary alcohol **6**. Reacting this compound with α-methylbenzylamine under palladium catalysis (*10*) provided the desired 2-amino substituted pyridine **7**. Reductive elimination of the tertiary hydroxyl group was accomplished with $NaBH_4$ in the presence of trifluoroacetic acid. *N*-Benzyl deprotection of the hydrochloride salt with hydrogen and Pd/C provided **1** in 66% yield. Thus using a 1,3-dipolar cycloaddition to provide a trisubstituted imidazole and functionalizing the 2-position had been demonstrated. The route provides the desired compound in 6 linear steps in 18% overall yield. The key step in this route, the TOSMIC cycloaddition to the *N*-methyl imine **4**, was of serious concern. First the reaction gave variable yields with some significant decomposition. In addition, DSC of the TOSMIC reagent indicated a 180 J/gm exotherm at 80 °C. Both of these factors led us to investigate alternate routes towards the synthesis of desired compound **1**.

Picolyl Anion Addition to Benzonitrile.

The next route involved the addition of the picolyl anion derived from **10** to 3-trifluoromethylbenzonitrile, and subsequent cyclization to the tetrasubstituted imidazole (Scheme 2).

This route initially appeared very attractive due to its convergency and the apparent ease with which the fragments could be prepared. 2-chloro-4-pyridine carboxaldehyde **8** (*11*) was reduced in the presence of methylamine to provide the desired secondary amine **9**. Acylation of amine **9** with isonipecotoyl chloride provided the amide **10** in 60% yield. Treatment of this compound with LiHMDS produced a deep blue solution of the picolyl anion. Treatment of that anion with 3-trifluoromethylbenzonitrile gave, after workup, the tetrasubstituted imidazole **11**. This intermediate was converted to the final product by palladium-catalyzed amination followed by Cbz deprotection in 85 and 90% yields respectively. This route provided the desired product **1** in five-steps from 2-chloro-4-pyridinecarboxaldehyde in 31% overall yield. Further investigation into the anion reaction revealed that several decomposition products were formed during the anion formation. Indirect evidence for anion decomposition was the isolation of *N*-methyl isonipecotamide from the reaction mixture. We attribute the formation of this side product to oxidative degradation of the picolyl anion to cleave the N-C bond. The picolyl degradation fragment was not observed. The apparent instability of this particular picolyl anion coupled with the variable yields of the addition/cyclization reaction led us to investigate other routes towards this target.

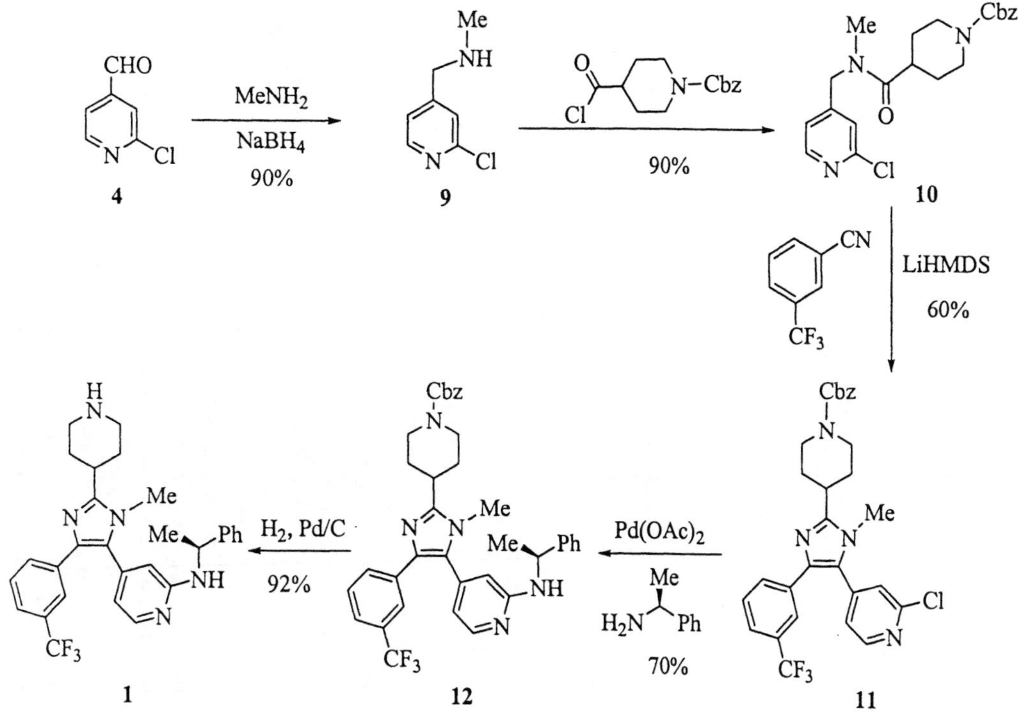

4 → 9: $MeNH_2$, $NaBH_4$, 90%

9 → 10: 1-Cbz-piperidine-4-carbonyl chloride, 90%

10 → 11: 3-(trifluoromethyl)benzonitrile, LiHMDS, 60%

11 → 12: $Pd(OAc)_2$, H_2N–CH(Me)Ph, 70%

12 → 1: H_2, Pd/C, 92%

Scheme 2. *Picolyl Anion Adition to Benzonitrile Approach to **1**.*

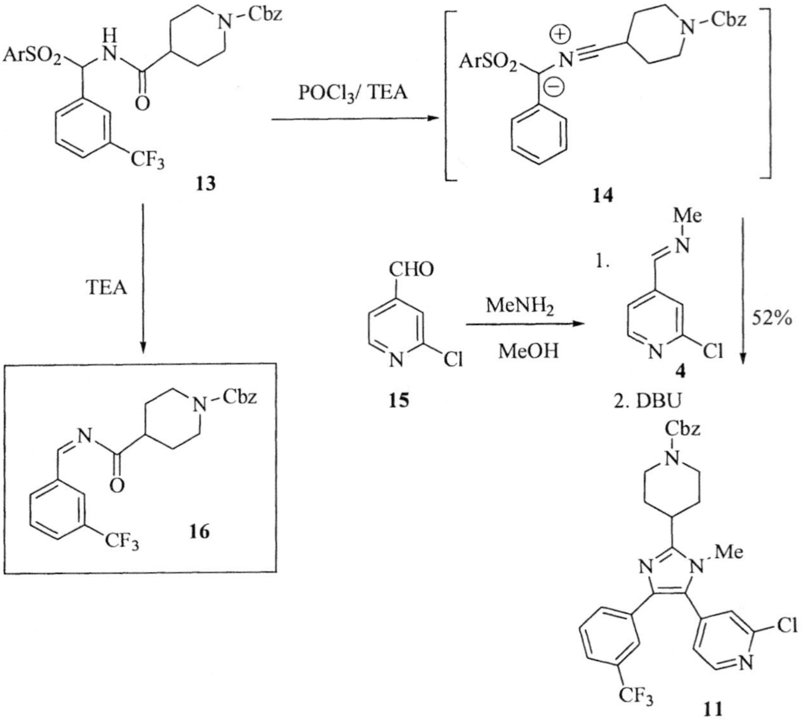

Scheme 3. Nitrilium Ylide Cyclization Approach.

1,3-Dipolar Cycloaddition of a Nitrilium Ylide.

We next turned our attention to 1,3-diolar cycloaddition reactions of nitrilium ylides as a direct approach towards the tetrasubstituted imidazole. Nitrilium ylides have been reported to undergo 1,3-dipolar cycloaddition with a number of dipolarophiles including imines.(*12-16*) Thus we reasoned that if the nitrilium ylide **14** could be formed, then reacted with the imine **4** derived from 2-chloro-pyridinecarboxaldehyde **15**, it might provide the desired imidazole (Scheme 3).

The tosyl amide precursor could be formed from 3-trifluoromethylbenzaldehyde, Cbz-isonipecotamide and toluenesulfinic acid following the procedure of Sisko.(*4, 5*) Exposing the amide **13** to $POCl_3$ in the presence of TEA gave a dark solution which was then treated with the *N*-methyl imine **4**. After warming to ambient temperature, the reaction was treated with DBU and allowed to stir overnight. The resulting mixture was chromatographed to provide the desired tetrasubstituted imidazole **11** in 52% yield. This reaction was examined further and it was found that elimination of toluenesulfinic acid to form the *N*-acylimine **16** was a major side reaction. Efforts to prevent this side reaction were never completely successful. As a result, we turned our attention towards taking advantage of the apparently facile formation of the *N*-acylimine **16**.

Acyl Anion Addition to Acyl Imines.

Previous work in synthesis of imidazoles has demonstrated that tetrasubstituted imidazoles could be formed from condensation of α-amidoketones (such as **17**, Scheme 4) with a primary amine. Thus, we envisioned that the desired imidazole (**11** or **12**) could arise from the reaction of the α-amidoketone **17** with methylamine. While α-amidoketone **17** may be prepared according to existing literature procedures; however, we envisioned the addition of acyl anion **18** to the *N*-acylimine **16** as a direct and more efficient route. Since we had just recently become aware of the propensity of tosyl amide **13** to eliminate sulfinic acid, we felt this may constitute a new entry into the synthesis of α-amidoketones.

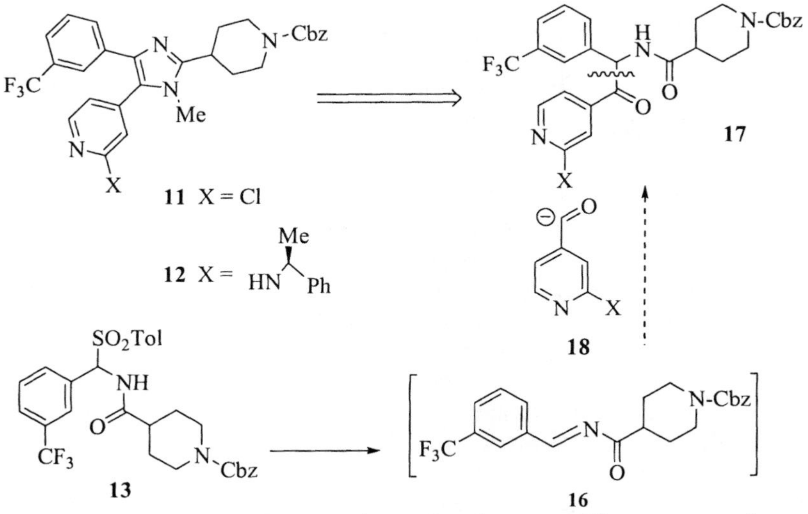

11 X = Cl

12 X = HN(CH(Me)Ph)

Scheme 4*. Retrosynthetic analysis of acyl anion addition approach.*

α-Amidoketones themselves are an important class of biologically important molecules.(*17-19*) Efforts to prepare diverse arrays of these compounds as enzyme inhibitors are current and extensive. In addition, these substrates represent a subclass of organic building blocks, which can be used to make stereochemically complex targets as well as important heterocycles such as imidazoles.(*20*)

In addition, the use of thiazolium catalyzed processes to prepare compounds which are the result of an acyl anion addition reaction have shown utility in synthetic organic chemistry. The benzoin condensation (*21-24*) and the Stetter reaction (*25-27*) represent two of the most powerful examples of these types of transformations (Figure 2.).

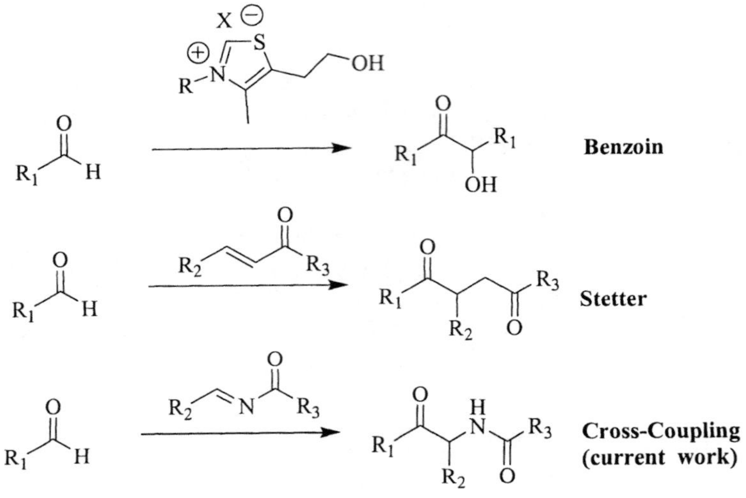

***Figure 2**: Thiazolium Catalyzed Acyl Anion Additions*

In order to expand the existing catalytic methodology towards the synthesis of α-amidoketones, we envisioned trapping the intermediate thiazolium stabilized acyl anion with an *N*-acylimine.(*28-30*) There are several potential problems with successfully executing this approach. Most importantly, the *N*-acylimine has to be sufficiently reactive to efficiently compete with another molecule of aldehyde (benzoin condensation), yet stable enough not to decompose under the reaction conditions or interfere with the thiazolium catalyst. Aryl sulfonylamides are stable, readily accessible substrates, which can undergo elimination of sulfinic acid to form *N*-acylimines under very mild conditions as previously discussed. We envisioned that by employing a tosyl amide in a reaction with an aldehyde, a thiazolium salt and a base such as triethylamine, we might be able to affect such a process. We were pleased to find that exposing tosyl amide **13** to a mixture of 2-chloro-4-pyridine-carboxaldehyde **15**, a commercially available thiazolium salt and triethylamine provided the desired α-amidoketone **17** in good yield (Scheme 5).

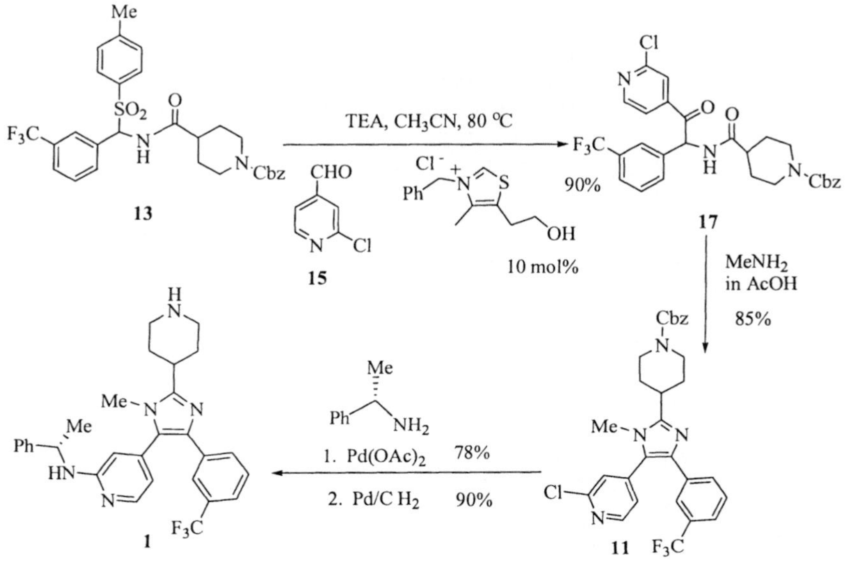

Scheme 5. Acyl anion approach to 1.

Reaction of **17** with $MeNH_2$ in EtOH in the presence of 1 equiv. of AcOH afforded the desired imidazole **11** in 85% yield. This intermediate was taken to the final compound as described earlier (Scheme 2).

This approach provides a four-step synthesis of **1** from 2-chloro-4-pyridinecarboxaldehyde in 60% overall yield.

Route Optimization

Evaluation of the four routes discussed above revealed that the acyl anion addition provided the most convergent approach to this compound and was selected for further development. While the route disclosed in Scheme 5 provides the desired product in four steps, a more convergent approach may arise from introducing the amine side chain earlier in the sequence (i.e. in the pyridine aldehyde). Thus to optimize this route we would need to have a practical synthesis of 2-[(1-phenylethyl)amino]-4-pyridinecarboxaldehyde **19**, optimize and understand the thiazolium catalyzed cross coupling process, and investigate the Cbz-deprotection and final salt formation

Aldehyde Synthesis

In search of a practical synthesis for 2-[(1-phenylethyl)amino]-4-pyridinecarboxaldehyde (**19**), we investigated three different approaches outlined in Scheme 6.

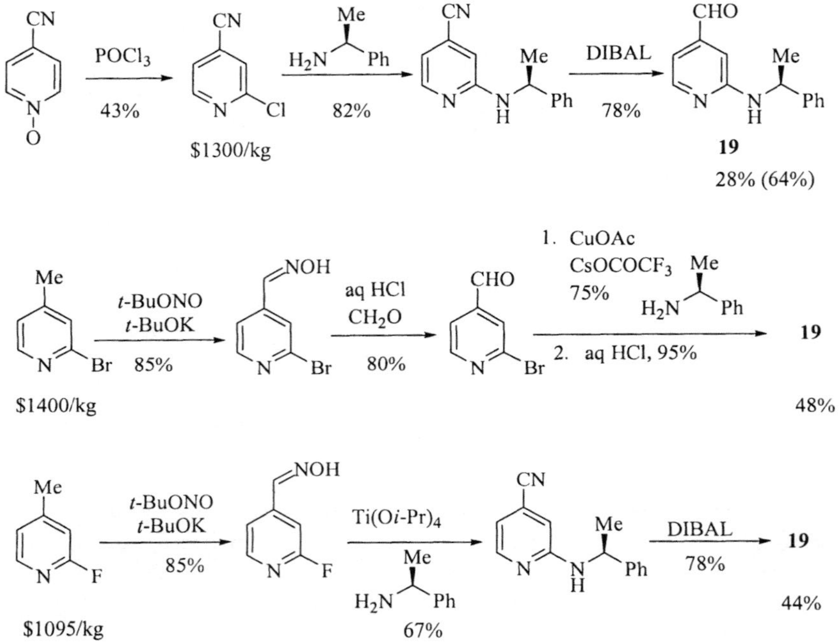

Scheme 6. *Aldehyde synthesis*

In the first approach, 4-cyanopyridine *N*-oxide was chlorinated with $POCl_3$ to give 2-chloro-4-cyanopyridine in 43% yield. Palladium catalyzed amination of 2-chloro-4-cyanopyridine with α-methyl benzylamine provided 2-(1-phenylethyl)amino-4-cyanopyridine in 85% yield . DIBAL reduction of the nitrile provided the desired amino aldehyde **19** in 78% yield. The aldehyde was isolated by quenching the DIBAL reaction mixture with aqueous acid. After discarding the toluene layer, the product was crystallized out as the bisulfite adduct by the slow addition of 20% aqueous sodium bisulfite solution. Two other routes were investigated involved oxidation of the corresponding picolines. Nitrosation of 2-bromo or 2-fluoro picoline with *t*-butyl nitrite in the presence of potassium *t*-butoxide provided the corresponding oximes in good yields. In the case of the bromo oxime, the oxime was hydrolyzed to provide the corresponding aldehyde, which was directly coupled with α-methyl benzylamine to provide the desired aldehyde **19**. Alternatively, the fluoro oxime was dehydrated to the nitrile during an amination promoted by titanium isopropoxide. The aminonitrile could be reduced with DIBAL as described before. Based on our evaluation of these routes, we selected the amination of the commercially available 2-chloro-4-cyanopyridine as the preferred sequence for the preparation of the desired aldehyde product **19**.

Optimization of the Coupling Reaction.

With the desired aldehyde in hand, we next turned our attention to optimizing the thiazolium-catalyzed coupling reaction. Exposing this aldehyde to a mixture of tosylamide **13**, thiazolium catalyst and TEA in THF provided the desired ketoamide **20** in 94% yield. Cyclization with *N*-methylamine as previously described provided imidazole **12** in 94% yield. Removal of the Cbz group provided compound **1**. This route provides the desired compound **1** in 5 steps and 56% overall yield from 2-chloro-4-cyanopyridine.

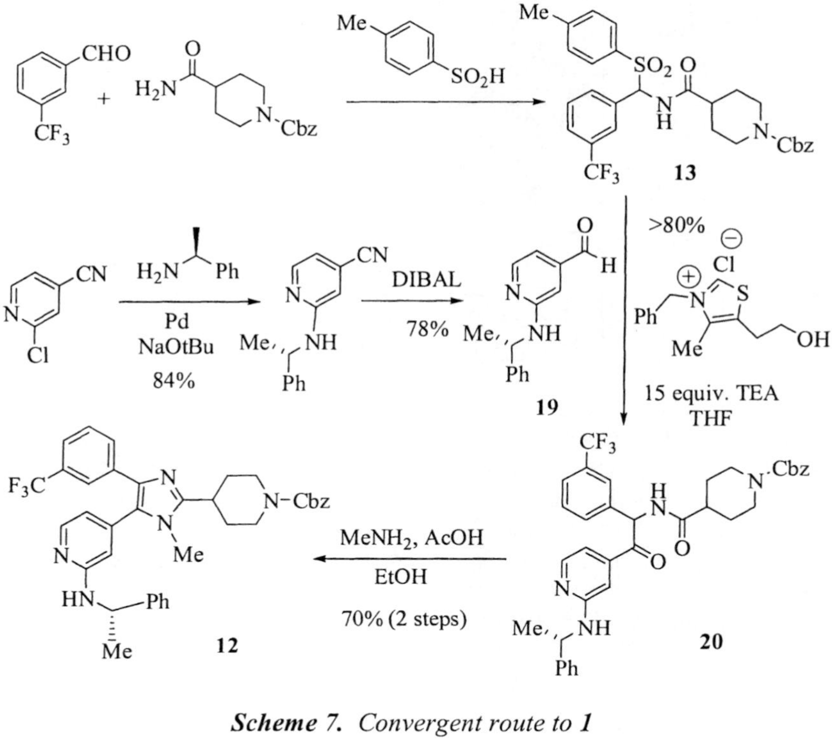

Scheme 7. Convergent route to 1

It became apparent that the optimal route towards the desired compound **1** would involve the thiazolium-catalyzed coupling reaction of aldehyde **19** with the tosyl amide **13**. In order to optimize this desired transformation, we decided to investigate the scope of this new reaction as well as understand its mechanistic underpinnings. The first parameter to investigate was the scope of reaction with respect to the aldehyde-coupling partner (Table 1.)

Table 1: Scope of reaction with respect to aldehyde

R_1CHO + Tol-SO_2-CH(Ph)-NH-C(O)-O*t*-Bu → (Et_3N, CH_2Cl_2, 35 °C; thiazolium catalyst: Bn-N⊕ thiazolium with CH₂CH₂OH, Cl⊖) → Ph-CH(NH-C(O)-O*t*-Bu)-C(O)-R_1

Entry	R_1	Time (h)	Yield (%)
1	4-pyridyl	0.25	85
2	Ph	24	75
3	2-Br-Ph	8	86
4	3-OMe-Ph	48	68
5	4-CN-Ph	0.25	80
6	2-furyl	24	73
7	3-pyridyl	24	93
8	CH_3	24	62
9	$BnOCH_2$	24	75
10	Ph-CH=CH	24	80

Inspection of the results listed in Table 1 reveals that electron poor aldehydes are better substrates than their electron rich counterparts. For example, while 3-methoxybenzaldehyde (entry 4) requires over 48 hours to react, 4-cyanobenzaldehyde reaction (entry 5) is complete in 15 minutes. Pyridyl aldehydes (entries 1 and 7) are excellent substrates for this reaction.

Examining the scope with respect to the amide group demonstrated that this had virtually no effect on the reaction (Table 2). Replacing the Boc group with Cbz or other amides did not affect the yield of the reaction (entry 2 *vs*. entry 3).

The biggest limitation of this methodology on the scope of the reaction came from the groups attached directly to the amide portion (R_2 and R_3). Alkyl groups are not tolerated at the R_2-position because of their ability to tautomerize to the corresponding enamide (entry 6). They are better tolerated at the R3-position (entries 1, 4 and 5). The formaldehyde variant (entry 7) could be used in this reaction although the reaction was sluggish. Varying the electronics of aryl groups in this position did not have a significant impact on the reaction yield or rate.

Table 2: Scope with respect to amide and imine

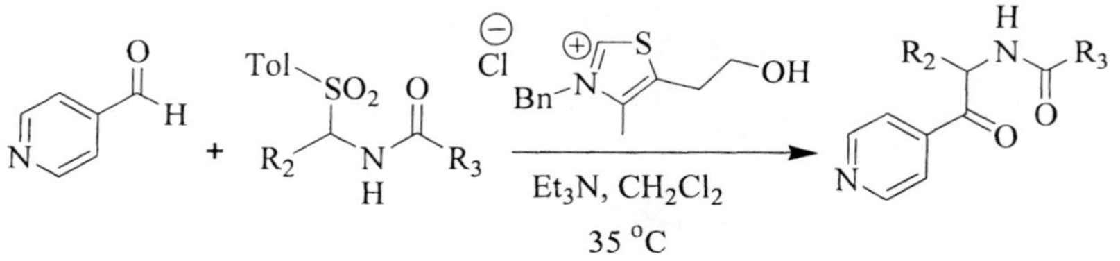

Et$_3$N, CH$_2$Cl$_2$ 35 °C

Entry	R_2	R_3	Time	Yield (%)
1	Ph	c-C_6H_{11}	30 min	98
2	Ph	OBn	15 min	96
3	Ph	O-t-Bu	15 min	85
4	4-F-Ph	c-C_6H_{11}	30 min	76
5	4-OMe-Ph	c-C_6H_{11}	30 min	84
6	c-C_6H_{11}	Ph	24 h	<10
7	H	Ph	24 h	58

It should also be noted that while methylene chloride and triethylamine were the optimum solvent and base for screening new substrates, other bases and solvents could be optimized for a particular substrate.

Mechanistic Information

To further understand the thiazolium-catalyzed cross coupling process, we conducted several studies to determine a possible mechanistic pathway for the reaction. A possible mechanism that may account for the outcome of this reaction is shown in Figure 3.

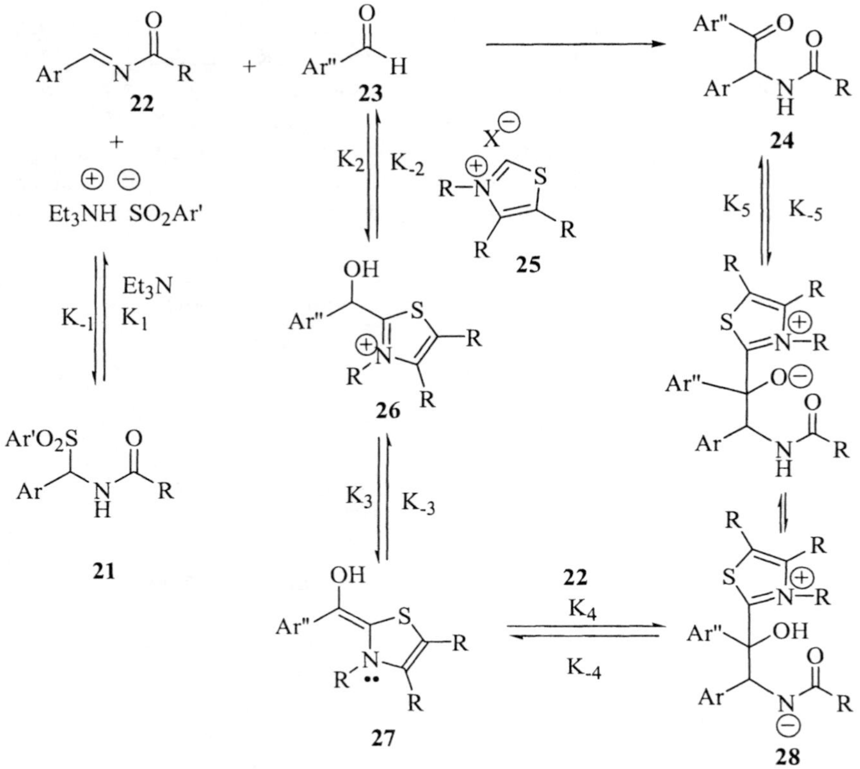

Ar N R O **22** + Ar" H O **23** → Ar" O Ar N H R O **24**

+ Et_3NH SO_2Ar' (⊕ ⊖)

K_{-1} Et_3N K_1

$Ar'O_2S$ Ar N H R O **21**

K_2 K_{-2} $X^{\ominus}$ R–N⊕ S R R **25**

OH Ar" S R ⊕N R R **26**

K_5 K_{-5} R S R ⊕N–R Ar" O⊖ Ar N H R O

K_3 K_{-3}

OH Ar" S R N R •• R **27**

22 K_4 K_{-4}

R S R ⊕N–R Ar" OH Ar N⊖ R O **28**

***Figure 3**. Mechanism 1*

NMR studies have revealed that under the reaction conditions, the tosyl amide **21** is in equilibrium with the *N*-acylimine **22**. The thiazolium salt **25** may be deprotonated by triethylamine then adds to the aldehyde **23** to provide the adduct **26**. Deprotonation provides the reactive enamine tautomer **27**. Reaction of this enamine with the *N*-acylimine **22** leads to the key C-C bond formation, which is followed by catalyst turnover to provide the desired product **24**.

A second alternate mechanism that cannot be ruled out and may be involved in this sequence is shown in Figure 4.

In this mechanism the addition of the thiazolium anion directly to the *N*-acylimine **22** is a lead process to form **29**, which is deprotonated to the enamine **30**. Addition of **30** with aldehyde **23** would provide the adduct **31** that is converted to the desired product **24**.

In order to differentiate between these two mechanisms and gain a better understanding of this process, we carried out a variety of experiments. The first experiment was to label the tosylamide with deuterium as illustrated in Scheme 8.

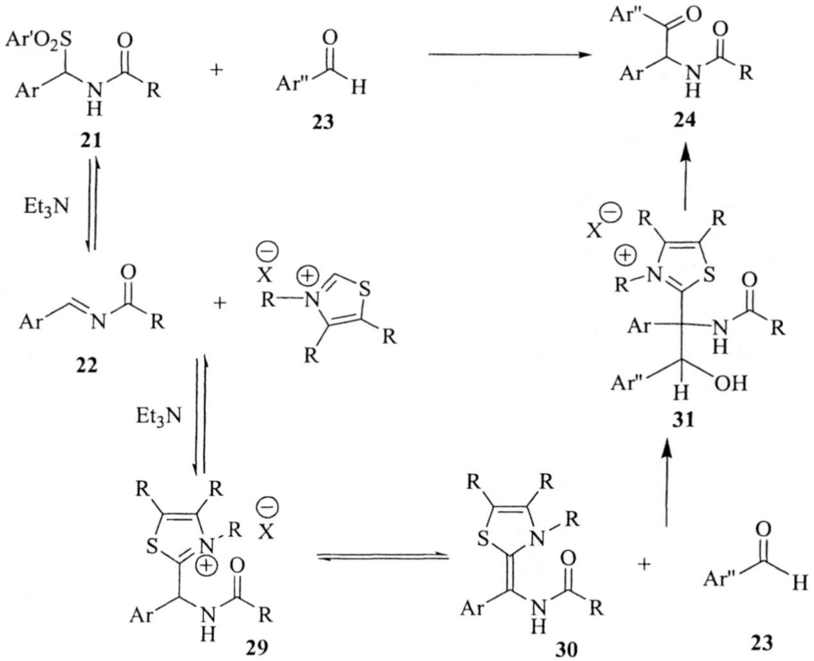

***Figure 4**: Mechanism 2*

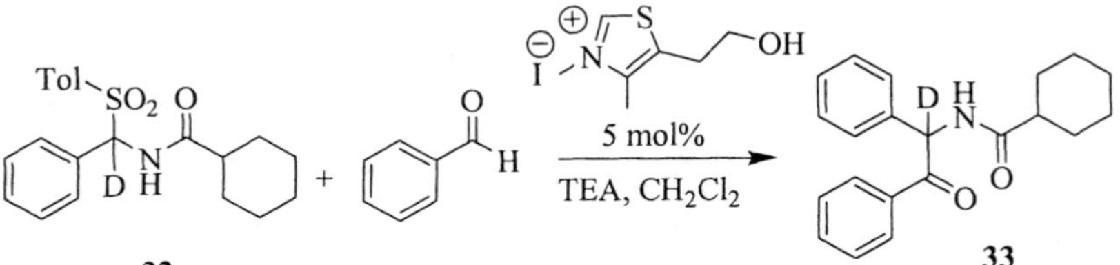

Tol SO2 O N D H 32 + O H; I⁻ N⁺ S OH 5 mol% TEA, CH_2Cl_2 → D H N O O 33

93% yield
>95% D incorporation

***Scheme 8**. Labeling Experiment*

If mechanism 1 was operating, then the deuterium should be maintained in the product. However, if mechanism 2 was operating, you would expect to not see any deuterium in the product. The deuterated tosyl amide **32** was easily prepared from commercially available deutero benzaldehyde and employed in the reaction. The product α-ketoamide **33** was isolated and characterized. The product was found to have >95% incorporation of the deuterium at the alpha position of the ketoamide. This rules out the possibility of the second mechanism and provides evidence that the product keto-amide **33** is not undergoing epimerization under the reaction conditions. Thus it should be possible to carry out an enantioselective version of this reactions utilizing the appropriate chiral thiazolium salt.

To gain further evidence for mechanism 1 we prepared the proposed intermediate aldehyde thiazole adduct **26** (Figure 1) by deprotonating thiazole with LDA and adding the resulting anion to benzaldehyde. Alkylation of this material with methyl iodide provided the corresponding salt, which was isolated and characterized. Employing a catalytic amount of this material in a reaction with a tosyl amide (**21**, Ar = Ar' = Ph, R = *c*-hexyl) and benzaldehyde provided the keto-amide product **24** in 83% yield.. Thus the thiazolium salt **26** is a viable intermediate in this reaction and an evidence that supports the first mechanism. (Scheme 9).

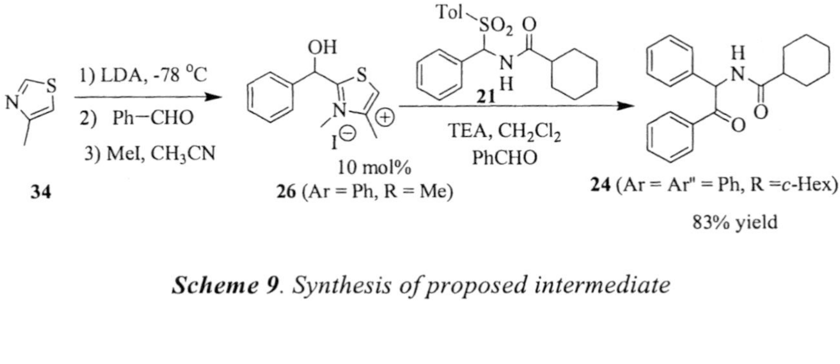

34: 1) LDA, -78 °C; 2) Ph−CHO; 3) MeI, CH_3CN → 26 (Ar = Ph, R = Me), 10 mol%; 21, TEA, CH_2Cl_2, PhCHO → **24** (Ar = Ar'' = Ph, R =*c*-Hex)

83% yield

***Scheme 9**. Synthesis of proposed intermediate*

In order to determine whether the reaction was under kinetic or thermodynamic control we performed a variety of crossover experiments as outlined in Scheme 10.

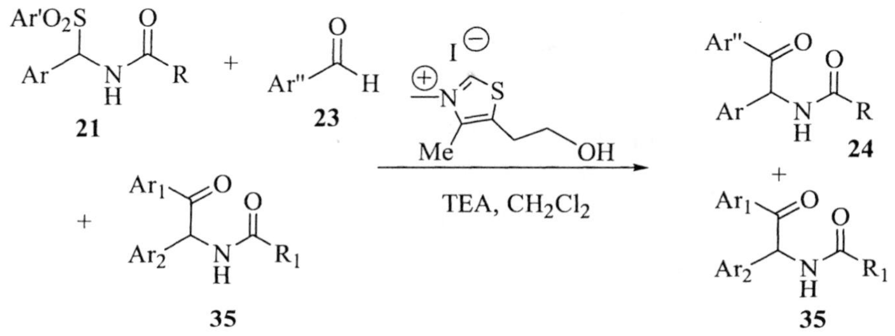

21 + 23 + 35 → (TEA, CH_2Cl_2) → 24 + 35

***Scheme 10**: Crossover experiments*

Carrying out the reaction of a distinct aldehyde (**23**) and tosylamide **21** in the presence of a completely differentiated keto-amide **35** provided only the product **24** from coupling of aldehyde **23** and tosyl amide **21** and not any of the components from the ketoamide **35** added at the start of the reaction. We performed this experiment several times and independently prepared samples of the crossover products and demonstrated that these were not present in these reactions.

Finally, in an attempt to determine the rate-determining step of this reaction, we measured the deuterium isotope effect of the aldehydic hydrogen. If the rate-determining event were addition of the thiazolium salt to the aldehyde, then one would expect to see an inverse isotope effect ($K_H/K_D < 1$). Conversely, if the proton transfer were the rate-determining event, then a large primary isotope effect would be observed. Finally, if C-C bond formation were in the rate-determining step, then there would not be an isotope effect since the label has been washed out at this point in the reaction sequence. Careful measurement of this parameter indicated that the K_H/K_D for benzaldehyde is 1.3. Because this number is neither <1 nor large, we conclude that this may be the result of a pre-equilibrium before the irreversible C-C bond forming event and that what we are observing is an equilibrium isotope effect. Another interpretation would be that the rate-determining event is addition of the thiazolium salt to the aldehyde and that the nature of the nucleophile (isoelectronic with carbene) may be responsible for the value of 1.3. Further experiments will be necessary to delineate the mechanistic factors of this reaction. A summary of our mechanistic evidence is presented schematically in Figure 5.

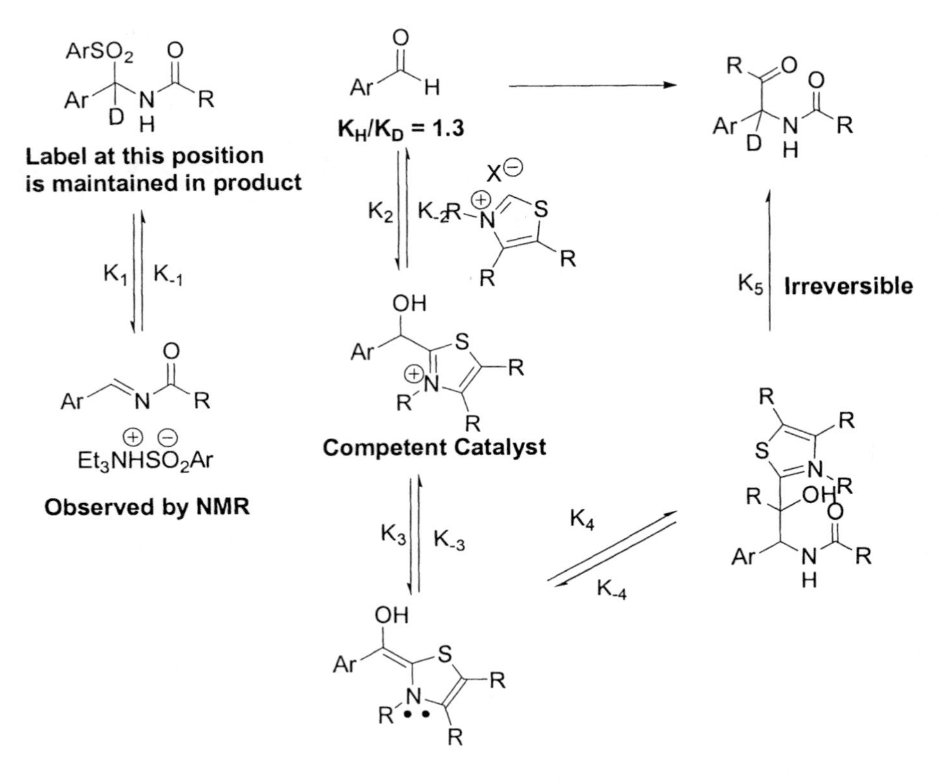

Figure 5. Mechanistic summary

Conclusion

In summary, we have disclosed new synthetic methodology to prepare a member of a class of tetrasubstituted imidazoles. The key step involves a convergent coupling step and is applicable to other members of this class of compounds. The intermediates necessary to prepare this compound were prepared using modifications to literature protocols. The pyridine fragment was prepared in 3 steps and 68% yield from 2-chloro-4-cyano pyridine. The tosylamide was prepared in two steps and 93% yield from isonipecotic acid. The key coupling step involves a thiazolium catalyzed cross acyloin-type condensation of an aldehyde with an *N*-acylimine (obtained from the tosylamide by *in situ* elimination of *p*-toluenesulfinic acid) to form an α-ketoamide. We have demonstrated the scope and some preliminary mechanistic studies concerning this new reaction. This α-ketoamide is then condensed with methyl ammonium acetate to provide the desired tetrasubstituted imidazole. Removal of theCbz group and formation of a pharmaceutically acceptable salt completes the synthesis in 6 steps and 38% overall yield.

References

1. Lee, J.C.; Laydon,J.T.; McDonnell, P.C.; Gallagher, T.F.; Kimar, S.; Green, D.; McNulty, D.; Blumenthal, M.J.; Heys, J.R.; Landvatter, S.W.; Strickler, J.E.; McLaughlin, M.M.; Siemens, I.R.; Fisher, S.M.; Livi, G.P.; White, J.R.; Adams, J.L.; Young, P.R. *Nature* **1994**, *372*, 739-745.
2. Adams, J.L.; Boehm, J.C.; Kassis, S.; Gorycki, P.D.; Webb, E.F.; Hall, R.; Sorenson, M.; Lee, J.C.; Ayrton, A.; Griswold, D.E.; Gallagher, T.F. *Bioorg. Med. Chem. Lett.* **1998,** *8*, 3111-3116.
3. For a recent review on the treatment of Inflammation by controlling cytokine pathways see Choy, E.H.S.; Panayi, G.S. *N. Engl. J. Med.* **2001**, *344*, 907-918.
4. Sisko, J.; Kassick, A.J.; Mellinger, M.; Filan, J.J.; Allen, A.; Olsen M. *J. Org. Chem.* **2000**, *65*, 1516-1524.
5. Sisko, J.; J. *Org. Chem.* **1998**; *63*, 4529-4531.
6. Claiborne, C.F.; Liverton, N.J.; Nguyen, K.T. *Tetrahedron Lett.* **1998**, 39, 8939-8942.
7. Liverton, N.L.; Butcher, J.W.; Claiborne, C.F.; Claremon, D.A.; Libby, B.E.; Nguyen, K.T.; Pitzenberger, S.M.; Selnick, H.G.; Smith, G.R.; Tebben, A.; Vacca, J.P.; Varga, S.L.; Agarwal, L.; Dancheck, K.; Forsyth, A.J.; Fletcher, D.S.; Frantz, B.; Hanlon, W.A.; Harper, C.F.; Hofsees, S.J.; Kostura, M.; Lin, J.; Luell, S.; O'Neill, E.A.; Orevillo, C.J.; Pang, M.; Parsons, J.; Rolando, A.; Sahly, Y.; Visco, D.M.; O'Keefe, S.J. *J. Med Chem* **1999**, *42*, 2180-2190.
8. Murry, J.A.; Frantz, D.E.; Soheili, A.; Tillyer, R.D.; Grabowski, E.J.J.; Reider, P.J. *J. Am. Chem. Soc.* **2001**, *123*, 9696-9697 .
9. van Leusen, A.M.; Wildemann, J.; Oldenziel, O.H. *J. Org. Chem.* **1977**, *42*, 1153.
10. Wagaw, S.; Buchwald, S.L. *J. Org. Chem.* **1996**, *6*1, 7240-7241.
11. Frey, L.; Marcantonio, K.; Frantz, D.E.; Murry, J.A.; Tillyer, R.D.; Grabowski, E.J.J.; Reider, P.J. *Tetrahedron Lett.* **2001**, *42*, 6815-6818.
12. Hansen, H.J.; Heimgertner, H. "Nitrile Ylides" in 1,3-Dipolar Cycloaddition Chemistry A. Padwa, Ed. 1984 John Wiley and Sons, Inc.
13. Padwa, A.; Carlsen, P.H.J. *J. Am Chem Soc.***1977**, *99*, 1514.
14. Padwa, Akamigata, N.;. *J. Am Chem Soc.***1977**, *99*, 1871.

15. Huisgen, R.; Stangl, H.; Sturm, H.J.; Wagenhofer, H. *Angew. Chem Int. Ed. Engl.* **1962**, *1,* 50.
16. Bunge, K.; Huisgen, R.; Raab, R.; Sturm, H.J. *Chem. Ber.* **1972,** *105*, 1307.
17. Lee, A.; Huang, L., Ellman *J. Am. Chem. Soc.* **1999**, *121*, 9907.
18. Rano, T.A.; Timkey, T.; Peterson, E.P.; Rotonda, J.; Nicholson, D.W.; Becker, J.W.; Chapman, K.T.; Thornberry, N.A. *Chem. Biol.* **1997**, *4,* 149-155.
19. Marquis, R.W.; Ru. Y.; Yamashita, D.S., Oh, H.J., Yen, J., Thompson, S.K., Carr, T.J., Levy, M.A., Tomaszek, T.A., Ijames, C.F., Smith, W.W., Zhao, B., Janson, C.A.; Abdel-Meguid, S.; D'Alessio, K.J.; McQueeney, M.S.; Veber, D.F. *Biorg. Med Chem* **1999**, *7*, 581-588 and *J. Am. Chem. Soc.* **1997**, *119*, 11351-11352.
20. Gupta, R.R.; Kumar, M.; Gupta, V. *Heterocyclic Chemistry II, Five membered Heterocycles*; Springer: Berlin, 1998.
21. Hassner, A.; Rai, K.M.L.*Comp. Org. Syn.* **1991**, *1*, 541-577.
22. Breslow, R. *J. Am. Chem. Soc.* **1958**, *8*0, 3719-3726.
23. Breslow, R.; Schmuck, C. *Tetrahedron Lett.* **1996**, *37*, 8241-8242.
24. Chen, Y.T.; Barletta, G.L.; Haghjoo, K.; Cheng, J.T.; Jordan, F. J. Org. Chem. 1994, 59, 7714-7722.
25. Stetter, H.; Kuhlman, H. *Chem. Ber.* **1976**, 2890-2896.
26. Stetter, H.; Kuhlman, H. *Synthesis.* **1975**, 379-380.
27. Stetter, H. *Angew. Chem. Int. Ed. Engl.* **1976**, *15*, 639-712.
28. Kinoshita, H.; Hayashi, Y.; Murata, Y.; Inomata, K. *Chem. Lett.* 1**993**, 1437.
29. Castells, J.; Lopez-Calahora, F.; Bassedas, M.; Urios, P. *Synthesis* **1988**, 314.
30. Katritzky, A.R.; Cheng, D.; Musgrave, R.P.*;* *Heterocycles*, **1996**, *42*, 1, 273.

Chapter 11

Solution Phase Synthesis of the Pulmonary Surfactant KL_4: A 21 Amino Acid Synthetic Protein

Ahmed F. Abdel-Magid[1], Mary Ellen Bos[1], Urs Eggmann[2], Cynthia A. Maryanoff[1], Lorraine Scott[1], Adrian Thaler[2], and Frank J. Villani[1]

[1]Johnson & Johnson Pharmaceutical Research and Development, Drug Evaluation–Chemical Development, Walsh and McKean Roads, Spring House, PA 19477
[2]Cilag AG, Hochstrasse 201, 8205 Schaffhausen, Switzerland

The synthetic 21-amino acid polypeptide KL_4 [$(KL_4)_4K$] was developed as a pulmonary surfactant to be used in treatment of respiratory distress syndrome (RDS) in infants (IRDS) and in adults (ARDS). KL_4 is designed to contain intermittent hydrophobic (Leucine) and hydrophilic (Lysine) regions to mimic the structure of the natural surfactant protein B (SP-B). The solution phase synthesis of KL_4 is a challenging task. The protection and deprotection patterns, the possibility of diastereomer formation in every coupling step, the solubility (or lack of it) of the different fragments and the isolation of intermediates are among the potential problems that face such a synthesis. The process research work resulted in a successful and efficient large scale synthesis of KL_4 which will be discussed.

Introduction

The epithelia of mammalian lungs are lined with endogenous pulmonary surfactants (PS), which facilitate breathing by aiding the transport of oxygen across the lung air-liquid interface. A deficiency in these surfactants is the primary cause of neonatal respiratory distress syndrome (RDS) and is also linked to RDS in adults. Native PS is a mixture of lipids and proteins, and although its exact composition is not well characterized, researchers have prepared a number of exogenous surfactants, such as KL_4 which are useful in the treatment of RDS in pre-term infants. KL_4 (**1**) is a 21-amino acid synthetic protein that was designed (*1*) to mimic the structure of one of the natural surfactant proteins known as surfactant protein B (SP-B). It contains intermittent hydrophobic (Leucine) and hydrophilic (Lysine) regions which are relevant to its mechanism of action.(*1*) KL_4 was initially developed at J&J as a pulmonary surfactant for the treatment of infants respiratory distress syndrome (IRDS). Later, the development was expanded to include treatment of adults (ARDS). This increased the demand for preparation of KL_4 to >10 kilogram quantities which was not easily achievable by traditional solid phase synthesis. To meet this demand, we needed to develop a solution phase synthesis that is suitable for large-scale production. In this paper, we report the results of our work on the solution phase synthesis of KL_4.

Different structural representations of KL_4 are illustrated in figure 1; due to the size of the intermediates and final product, the structures reported herein will be represented by the letter designations **K** and **L** for lysine and leucine respectively.(*2*)

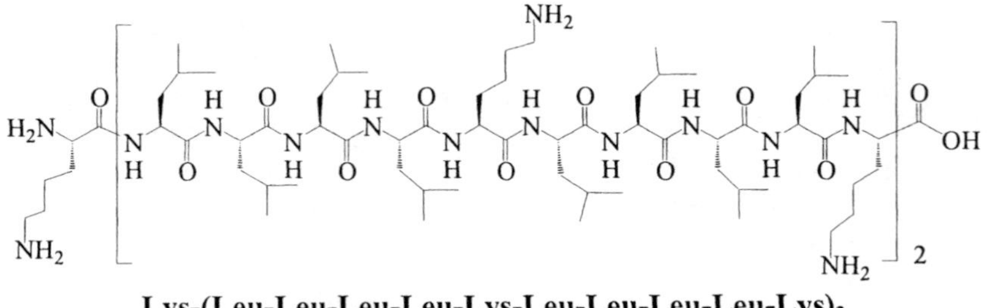

Lys-(Leu-Leu-Leu-Leu-Lys-Leu-Leu-Leu-Leu-Lys)$_2$

K(LLLLKLLLLK)$_2$; K(LLLLK)$_4$; (KLLLL)$_4$K or (KL$_4$)$_4$K

(KL$_4$)

(1)

***Figure 1**: Different representations of KL_4*

Results and Discussion:

Although, KL_4 consists of repeat units of Lysine (Lys or K) and Leucine (Leu or L) residues (Lys-Leu-Leu-LeLeu or KLLLL), the solution phase synthesis proved to be challenging. The choice of suitable building fragments, the protection and deprotection patterns, the possibility of epimerization and diastereomer formation in every coupling step, the decreased solubility of the fully protected large fragments, the work up conditions and isolation and purification of intermediates were among the problems that faced the solution phase synthesis. We anticipated that fully protected polypeptide intermediates to become less soluble as the length of the chain increases. This would cause chemical reactions to be sluggish and unpredictable. Additionally, the possible increased epimerization at the activated C-terminal amino acid of the polypeptide fragment in the coupling steps and the difficulty in purifying the insoluble protected peptides are consequential problems.(*3*) The first generation synthesis was planned as a linear process (Scheme 1). The building fragments were carefully chosen to minimize the use of terminal Lysine free acid components in coupling reactions. That was based on literature reports, which show more side reactions and epimerization at the α-carbon, when polypeptides containing this acid are used in coupling reactions.(*4*) The same report shows that terminal leucine acid residues give the least epimerization side reactions. As a result, the following three fragments were chosen as building units:

(a) **Start Fragment**: H-LLK(Z)-OBn (**2**) consists of three amino acid residues representing amino acid residues 21-19 in the final KL_4 chain (Scheme 1).

(b) **Intermediate Fragment**: Boc-LLK(Z)LL-OH (**3**), consists of five amino acid residues, to be used three times to add residues 18-14, 13-9, and 8-4 successively (Scheme 1).

(c) **End Fragment**: Z-K(Z)LL-OH (**4**), consists of three amino acid residues, representing residues 3-1 (Scheme 1).

The protection strategy was designed to use *tert*-butoxycarbonyl (Boc) protective groups for the α-amino groups of leucine and benzyloxycarbonyl (Cbz or Z) protective groups for the side chain amines of lysine as well as the α-amino group on the final lysine fragment. The terminal carboxy group of lysine was protected as benzyl (Bn) ester. This choice of protective groups allows the selective removal of the Boc-groups from α-amines for the chain propagation. The Z-groups together with the terminal benzyl ester are removed simultaneously at the end of synthesis in one deprotection process, namely, hydrogenation. The syntheses of the three building units are illustrated in Schemes 2, 3 and 4. All three fragments were prepared from the same starting material Boc-LL-OCH_3 (**5**). The methyl ester **5** was prepared in 90% yield by coupling Boc-L-OH with H-L-OMe in EtOAc using DCC/HOBt combination as the coupling agent.(*5*)

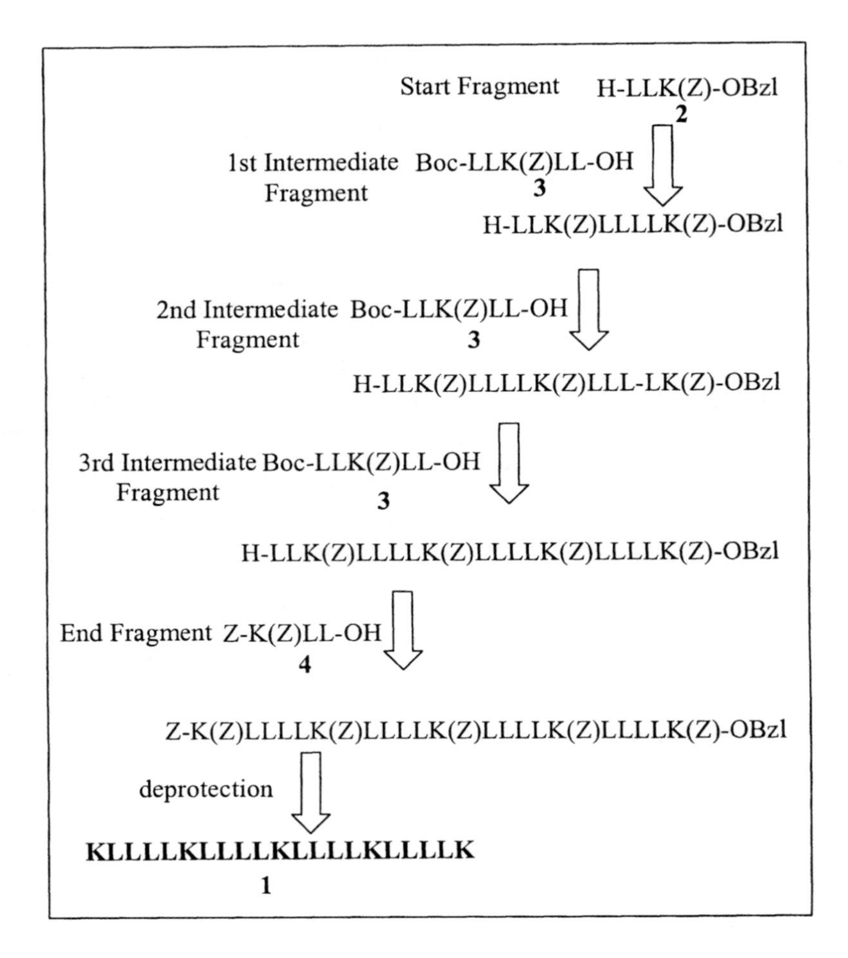

***Scheme 1**: Synthesis Plan*

Synthesis of the Start Fragment 2:

Hydrolysis of ester **5** was accomplished with NaOH in aqueous acetone at rt to give Boc-LL-OH (**6**) in 92% yield after acidification. The acid **6** was coupled to H-K(Z)-OBn using DCC/HOBt in DMF to give the tripeptide **7** in 93% isolated yield. The Boc-group was removed using anhydrous HCl in EtOAc to give HCl • H-LL-K(Z)-OBn (**2**) in 92% and about 77% overall yield from **5**.

Boc-LL-OCH_3 (5) —[NaOH, acetone/H_2O; 89.6-91.7%]→ Boc-LL-OH (6) —[H-K(Z)-OBn; DCC, HOBt/DMF; 93.4%]→

Boc-LLK(Z)-OBn (7) —[HCl/EtOAc; 92.1%]→ HCl • H-LLK(Z)-OBn (2)
start fragment (carboxy terminus)

***Scheme 2**: Synthesis of start fragment 2*

Synthesis of the Intermediate Fragment 3:

Removal of the Boc group of **5** with anhydrous HCl in EtOAc gives the corresponding amine salt **8**. Coupling of amine **8** with the mixed anhydride made from Boc-K(Z)-OH and *iso*-butyl chloroformate in the presence of 4-methyl morpholine (NMM) gave the tripeptide **9** in 90% yield. Removal of the Boc protective group was carried out with anhydrous HCl in EtOAc to give **10** which is coupled to **6** using DIC/HOBt in DMF to give **11**. The latter was hydrolyzed with NaOH in aqueous acetone to give the intermediate fragment **3** in about 66% overall yield starting from **5**.

Boc-LL-OCH_3 (5) —[HCl/EtOAc; 96%]→ HCl • H-LL-OCH_3 (8) —[Boc-K(Z)-OH; $ClCO_2i$-Bu, NMM; 90%]→

Boc-K(Z)LL-OCH_3 (9) —[HCl/EtOAc; 91%]→ HCl • H-K(Z)LL-OCH_3 (10) —[Boc-LL-OH (**6**); DIC, HOBt, DMF; 86%]→

Boc-LLK(Z)LL-OCH_3 (11) —[NaOH, acetone/H_2O; 98%]→ Boc-LLK(Z)LL-OH (3)
intermediate fragment

***Scheme 3**: Synthesis of intermediate fragment 3*

Synthesis of the End Fragment 4:

The end fragment **4** was obtained in about 86% yield from **5** (Scheme 4) via removal of the Boc-protective group of the amine with anhydrous HCl in EtOAc, followed by coupling with Z-Lys(Z)-OH (**12**) and hydrolysis of the resulting ester **13** with LiOH in aq. Acetone.

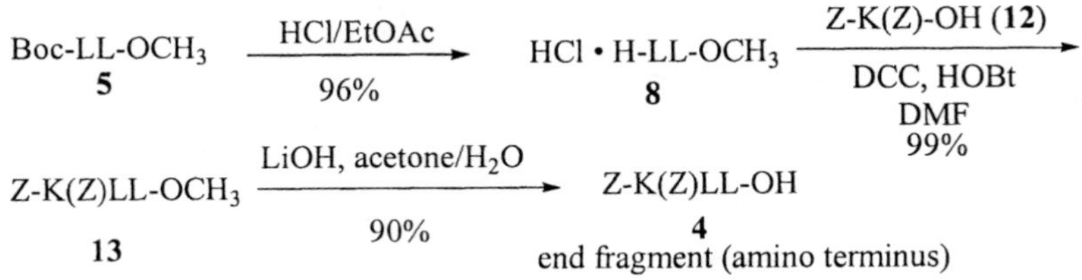

$$\underset{\mathbf{5}}{\text{Boc-LL-OCH}_3} \xrightarrow[96\%]{\text{HCl/EtOAc}} \underset{\mathbf{8}}{\text{HCl} \bullet \text{H-LL-OCH}_3} \xrightarrow[\text{DCC, HOBt, DMF, } 99\%]{\text{Z-K(Z)-OH } (\mathbf{12})}$$

$$\underset{\mathbf{13}}{\text{Z-K(Z)LL-OCH}_3} \xrightarrow[90\%]{\text{LiOH, acetone/H}_2\text{O}} \underset{\mathbf{4}}{\text{Z-K(Z)LL-OH}}$$

end fragment (amino terminus)

Scheme 4: *Synthesis of the end fragment* ***4***

Fragment Couplings: First Synthesis.

With all three building fragments in hand, we continued the synthetic sequence as outlined in Scheme 5. We decided, following a preliminary run through the sequence, to carry out the sequential steps without purification of intermediates and perform one purification by chromatography on the final product. This was mainly due to the low solubility of the fully-protected intermediates and their unique physical properties that excluded recrystallization as an option. A considerable effort was made to optimize step for yield and purity to compensate for the lack of purifications during the synthesis.

Coupling of the start fragment **2** with the first intermediate fragment **3** was carried out under different coupling conditions to give the product **14** (Boc-LLK(Z)LLLLK(Z)-OBn or Boc-14-21-OBn) (*2*) which contains the eight amino acid residues number 14-21. The optimum results were obtained using the combination of HBTU/HOBt/DIPEA in DMF/CH_3CN solvent mixture to give the product **14** in 93% isolated yield with <0.7% of the *D*-18 diastereomer Boc-LLK(Z)LL*****LLK(Z)-OBn.(*6*) Removal of the Boc-protective group was achieved in high yield by treatment of Boc-14-21-OBn (**14**) with excess anhydrous HCl in EtOAc at -15 to -10 °C. The work up required removal of the large excess of HCl by multiple dilutions with petroleum ether and subsequent evaporations at low temperature (under vacuum) to finally precipitate the hydrochloride salt of the amine product, HCl • H-LLK(Z)LLLLK(Z)-OBn (**15a**). A more convenient deprotection procedure was developed using trifluoroacetic acid (TFA). The use of TFA is not usually preferred in large scale reactions because of the need to evaporate most of the TFA at the end of the reaction. However, stirring **14** in TFA (5 mL/g) at 0 °C for 1h and pouring the resulting solution onto ice/water, precipitated the TFA salt TFA • H-LLK(Z)LLLLK(Z)-OBn (**15b**) as a white solid which was simply collected after filtration and washing away the excess acid with water.

Either the HCl **15a** or the TFA salt **15b** is used in the following coupling step with the second intermediate fragment **3**. The HCl salt **15a** is the least soluble of the two salts and was initially used in this coupling step. As mentioned earlier, the decreased solubility of the larger fragments was an anticipated complication and conditions needed to be modified to deal with such a problem. Thus, after some experimentation, a procedure was developed which included dissolving **15a** in DMF/1-methyl-2-pyrrolidone (NMP)/THF solvent mixture (8:3:5 ratio) with lithium perchlorate as an additive (7.5 mole %) to enhance solubility and reduce gelling. This was followed by addition of **3** then (i-Pr)$_2$NEt, HBTU and HOOBt at -5 °C. After stirring at rt for 5-6 h, the THF was distilled off and the remaining residue was added slowly to 1% aq. $NaHCO_3$ solution and the solid was collected by filtration. The crude yellow solid was then triturated with hot acetone/MTBE (3:1) to remove the yellow color and give the product Boc-LLK(Z)LLLLK(Z)-LLLLK(Z)-OBn (**16**) in 88-94% yield and 92-96% purity with 1-2% of the *D*-13 diastereomer Boc-LLK(Z)LL***LLK(Z)-LLLLK(Z)-OBn.(*6*) The TFA salt,(**15b**) is more soluble in DMF than the HCl salt **15a**. This allowed us to carry out the coupling reaction in less volume of one solvent (DMF) and without additives. This also enabled us to examine other coupling conditions. The results from several reactions under different coupling conditions are listed in Table I. All reaction conditions gave the *D*-13 diastereomer in variable amounts. Conditions b, (entries 2, 5 & 8, HBTU, HOOBt, (i-Pr)$_2$NEt, at -5 °C, 5 h) gave the least amounts of the *D*-13 diastereomer, 0.14-0.4%. Conditions f (entry 7), on the other hand, gave the highest ratio of this diastereomer, 14.9%.

Table I: Coupling of 15b with 3 in DMF

entry	*Reaction Conditions*[1]	*Scale (mmol)*	*Yield (%)*	*Area % Purity*	*D-13 Diast. (Area %)*
1	a	1.4	97	84.4	3.30
2	b	1.4	90	94.4	0.14
3	c	1.4	84	94.2	1.20
4	d	1.4	97	81.2	0.70
5	b	13.6	96	93.0	0.40
6	e	1.4	95	89.7	4.40
7	f	1.4	96	79.8	14.9
8	b	25.0	95	88.0	0.40

1. Conditions: (a) DIC, HOBt, NMM, rt, overnight. (b) HBTU, HOOBt, (i-Pr)$_2$NEt, -5 °C, 5 h. (c) HBTU, HOBt, (i-Pr)$_2$NEt, -4 °C, 5 h. (d) DIC, HOOBt, NMM, rt, overnight. (e) WSCDI, HOOBt, (i-Pr)$_2$ NEt, -4 °C - rt. (f) WSCDI, HOBt, (i-Pr)$_2$NEt, -4 °C - rt.

The work up, in all cases, consisted of pouring the reaction mixture into ice water and stirring at rt. Addition of aqueous sat. NaCl or sat. $NaHCO_3$ to the aqueous mixture prevented emulsion formation and gave an easy to filter solid. The solid was collected by filtration and the crude product was purified by trituration in hot acetone/MTBE (3:1) or in MeOH/H_2O (1:1). Both salts, **15a** and **15b**, enabled us to prepare the desired product, Boc-LLK(Z)LLLLK(Z)LLLLK(Z)-OBn (Boc-9-21-OBn, **16**) in excellent yield and high purity.

Removal of the Boc-protective group of **16** was achieved by treatment with excess anhydrous HCl in EtOAc or with neat TFA. Thus, **16** was treated with large excess of anhydrous HCl in ethyl acetate at -16 to -14 °C for 1h followed by removal of the excess HCl under reduced pressure at -20 to 0 °C. The solution was diluted with diethyl ether at -3 to 3 °C and the pH of the mixture was adjusted to 1.5-2.5 by addition aqueous potassium bicarbonate. The resulting mixture was warmed to 19 to 23 °C and stirred for 30 min. The solid product was collected by filtration, rinsed with H_2O and dried *in vacuo* at 35 to 40 °C to give HCl • H-LLK(Z)LLLLK(Z)LLLLK(Z)-OBzl (**17a**) in 95% yield and 87.7 area % purity.

Alternatively, compound **16** was treated with TFA at 0 °C for 1h then poured into water with efficient stirring to precipitate the desired product, TFA • H-LLK(Z)LLLLK(Z)LLLLK(Z)-OBzl (**17b**) in 90-97% yield. The HPLC purity of the product was consistently higher than that obtained using HCl, ranging from 90-92%, (Table II) however, the isolation of the product was not easy. Adding the TFA solution to ice water results in formation of a sticky semi-solid product. On the laboratory scale we decant the aqueous liquid and treat the residue with hot MeOH/H_2O (about 95:5) to obtain a solid product. We decided however to use the HCl salt since it was relatively easier to isolate during large scale production.

Table II: Removal of Boc group of 16

entry	*Reagent*	*Scale (mmol)*	*Yield (%)*	*Area% purity (HPLC)*
1	HCl/EtOAc	17.9	95	87.7
2	TFA	0.5	95	91.2
3	TFA	1.4	97	90.4
4	TFA	2.0	93	92.1
5	TFA	2.8	97	90.3

The coupling of H-9-21-OBn, (**17**) and Boc-LLK(Z)LL-OH, (**3**) was the most problematic in the entire sequence. The greatest problem was caused by the very poor solubility of **17a** in most solvents. Even at very low concentrations, thick gels were formed which made it difficult to stir and led to poor coupling results. Longer reaction times resulted in the formation of thick gelatinous layers that stuck to the inner walls of the flask (or reactor) and practically gave no reaction. Addition of some lithium salts (e.g., $LiBF_4$) reduced the gelling problem but didn't eliminate it. Other additives such as dimethylpropylene urea, tetramethylurea, higher alcohols, etc. had some variable effects in reducing the gelling but none gave satisfactory results. The most surprising effect came from adding water to the coupling reaction. We found that addition of small amounts of water remarkably reduced the gelling and allowed the reaction to proceed faster to completion and gave reproducible results. Addition of water to coupling reactions is not a common practice and in most cases it means the demise of such a reaction. We presume that water decreases the intermolecular interactions leading to formation of β-sheets or other transverse networks that increase viscosity and cause gelling. Consequently that improves the ability to stir and increases the accessibility (of reagents) to the peptide chain. The reaction was carried out in THF/ NMP/water (7:2.5:1) in the presence of $LiBF_4$. The mixture was first homogenized (using a high-shear mixer) for a few minutes followed by the addition of H_2O and the coupling agents (HBTU, HOOBt and DIPEA) then stirring the mixture at rt to give the desired product **18** (Boc-LLK(Z)LLLLK(Z)LLLLK(Z)LLLLK(Z)-OBn or Boc-4-21-OBn) in 82-89% yield and HPLC area% purity of 81-89%. The product typically contained 4-6% of (Boc-LLK(Z)L**L***LLK(Z)LLLLK(Z)LLLLK(Z)-OBn), the *D*-8 diasteromeric impurity.(*6*)

The removal of the Boc-protective group in **18** was achieved using excess HCl in EtOAc as previously described for compounds **14** and **16** to give the product **19a**. The coupling of **19a** with the end fragment Z-K(Z)LL-OH (**4**) was relatively easier to carry out compared to the previous coupling step. As in the previous coupling step, the addition of water and $LiBF_4$ was necessary for the success of the reaction; the concentration could be doubled and volume efficiency was thus increased. The use of DIC as a coupling agent gave less epimerization than HBTU. The product of this coupling is the fully protected KL_4 (Z-K(Z)LLLLK(Z)LLLLK(Z)LLLLK(Z)LLLLK(Z)-OBn, **20**) which was isolated in 96% yield and HPLC area% purity of about 80% containing about 4% of the *D*-3 epimer Z-K(Z)L**L***LLK(Z)LLLLK(Z)LLLLK(Z)LLLLK(Z)-OBn.(*6*)

Alternatively, the removal of the Boc-protective group in **18** was achieved using TFA. The product (**19b**) was isolated in near quantitative yield and in 88% HPLC area% purity. The use of **19b** greatly improved the final coupling step. The TFA salt is much more soluble, gel-formation was not observed and the coupling step didn't require any additives. The use of HBTU/HOOBt/DIPEA gave the desired product **20** in 92% yield and HPLC purity of 90%. Most importantly, the ratio of epimerization to the *D*-3 diastereomer (*6*) was much lower than the previous conditions, down to 1.1% compared to 4% when the HCl salt was used.

Boc-LLK(Z)LL-OH **3** + HCl • H-LLK(Z)-OBzl **2** —(a)→ Boc-LLK(Z)LLLLK(Z)-OBzl **14**

—(b) or (c)→ HX • H-LLK(Z)LLLLK(Z)-OBzl (**15a**: X = Cl; **15b**: X = $OCOCF_3$) —Boc-LLK(Z)LL-OH **3**, (d) or (e)→

Boc-LLK(Z)LLLLK(Z)LLLLK(Z)-OBzl **16** —(b)→ HCl • H-LLK(Z)LLLLK(Z)LLLLK(Z)-OBzl **17a**

—Boc-LLK(Z)LL-OH **3**, (f)→ Boc-LLK(Z)LLLLK(Z)LLLLK(Z)LLLLK(Z)-OBzl **18** —(b) or (c)→

HX • H-LLK(Z)LLLLK(Z)LLLLK(Z)LLLLK(Z)-OBzl (**19a**: X = Cl; **19b**: X = $OCOCF_3$) —Z-K(Z)LL-OH **4**, (g) or (h)→

Z-(K(Z)LLLL)$_4$K(Z)-OBzl **20** —(i)→ H-KLLLLKLLLLKLLLLKLLLLK-OH [(KLLLL)$_4$K] **1**

Reagents; yields: (a) HBTU, HOBt, DIPEA, DMF/CH_3CN; 0 °C, 93%, (b) HCl/EtOAc, -15 to -20 °C (c) TFA, 0 °C (d) HBTU, HOOBt, DIPEA, $LiClO_4$, DMF/NMP, -5 °C to rt, 89% (e) HBTU, HOOBt, DIPEA, DMF, -5 °C; 96%, (f) HBTU, HOOBt, DIPEA, $LiBF_4$, NMP/THF/H_2O; 89%, (g) DIC, HOOBt, DIPEA, $LiBF_4$, NMP/THF 20-25 °C, 96% (h) HBTU, HOOBt, DIPEA, NMP/THF -9 °C to rt; 92%, (i) H_2, Pd/C, TFA/AcOH/H_2O; 95%.

***Scheme 5**: Fragment Couplings (Synthesis of KL-4)*

The final step in the synthesis was the global deprotection. Removal of all the Z-groups on the side chain amino groups of the lysine residues as well as the cleavage of the terminal benzyl ester was accomplished by catalytic hydrogenation over palladium in acidic solution. The use of TFA/AcOH/H_2O solvent mixture gave the best results. The deprotection was complete in about 3h and the crude product was isolated after filtration to remove the catalyst and evaporation of the solvent. It is noteworthy that the removal of the catalyst should be done in the absence of air (oxygen) to eliminate, or minimize, contamination of the product with palladium. The pure product was isolated by prep HPLC as the acetate salt.

Convergent Synthesis of KL_4

In the latter steps of the above synthesis several problems resulted from the lower solubility of intermediates, including low reaction throughput (from high dilution) and higher levels of epimerization that produced undesired diastereomeric products. In addition, purification of products was not attainable and the impurities were carried through to the final product which was purified by chromatography. In spite of these problems, the synthesis sequence was carried out successfully to prepare kilogram quantities of the desired product, KL_4 acetate.

Since most of the problems encountered with the synthesis were the result of the physical properties of the larger peptide fragments rather than the chemistry used, a convergent synthesis in which two key intermediates are used, would solve most of the problems associated with solubility of large fragments. However, a potential synthetic problem with this approach would be the preparation of a peptide chain with a free carboxy terminus, which is non-conventional. The new synthesis would add minimum changes to the existing first synthesis thus minimizing analytical method development. It would also give a better in-process control with two key intermediates and one critical coupling procedure. Thus, the synthesis was designed to use the 13-residue intermediate (**17**), obtained in high purity in the first synthesis, as the amine (representing residues 9-21). An acid intermediate, Z-K(Z)LLLLK(Z)LL-OMe (**21**) representing the 1-8 residues, was synthesized to test this approach. The synthesis of this new intermediate was accomplished by coupling of the 5-residue intermediate fragment H-LLK(Z)LL-OMe (**3**) and the 3-residue end fragment Z-K(Z)LL-OH (**4**) (Scheme 6). The resulting ester Z-K(Z)LLLLK(Z)LL-OMe (Z-1/8-OMe) (**21**) was not stable under most basic hydrolysis conditions. Attempted hydrolysis of the methyl ester using LiOH, NaOH, Cs_2CO_3, $KOSiMe_3$ caused unexpected decomposition and formation of complex reaction mixtures. We initially attributed these results, in part, to the poor solubility of Z-1/8-OMe in typical hydrolysis reaction mixtures, however, in THF/MeOH with aqueous NaOH as the base the reaction mixture is nearly homogeneous but the results were still uniformly poor.

Z-K(Z)LL-OH (**4**) —[HCl • H-LLK(Z)LL-OMe; HBTU/HOOBt, DIPEA/DMF, -4 °C]→ Z-K(Z)LLLLK(Z)LL-OMe (**21**)

—[ester hydrolysis]—X→ Z-K(Z)LLLLK(Z)LL-OH (**22**)

***Scheme 6**: Attempted synthesis of Z-K(Z)LLLLK(Z)LL-OH*

An alternative 3-residue end fragment (Boc-K(Z)LL-OH) (**23**), obtained by hydrolysis of ester **9**, was used to prepare Boc-K(Z)LLLLK(Z)LL-OMe (**24**) (Scheme 6). This minor structural change dramatically improved the stability of the intermediate under basic conditions. The hydrolysis of **24** was not accompanied by decomposition and was carried out using several procedures. Traditional hydrolysis conditions such as LiOH or NaOH in acetone, THF or MeOH resulted in poor reactions and high levels of epimerization (10-15%) at the α-carbon of the ester. The epimerization at the α-carbon of an ester is a major concern in the hydrolysis of peptide esters under basic reaction conditions. This is even more of a problem with larger fully protected polypeptide esters that are usually sparingly soluble in solvents such as methanol and acetone, commonly used in these reactions with an alkali hydroxide such as NaOH or LiOH. This poor solubility results in slow hydrolysis and increased epimerization at the α-carbon of the ester. A better result was obtained from hydrolysis of **24** with NaOH in DMF, which gave the product Boc-K(Z)LLLLK(Z)LL-OH (**25**) in good chemical yield with about 5-6% of Boc-K(Z)LLLLK(Z)LL^*-OH, the diastereomer resulting from the epimerization at the α-carbon. In spite of this relatively high level of epimerization, the hydrolysis with NaOH/DMF was a better alternative and offered a great improvement over other procedures.

Our continued search for better alternatives to traditional saponification procedures led to the use of tetraalkylammonium hydroxides. Tetraalkylammonium hydroxides are very strong bases and have the advantage of being soluble in water and in many organic solvents. However, these bases, except for some scattered reports, were not fully utilized in carboxylic ester hydrolysis, particularly for polypeptide esters. The use of tetraalkylammonium hydroxides in hydrolysis of small easily soluble polypeptide esters does not show much of an advantage over the traditional conditions such as NaOH in aqueous acetone. However, fully-protected larger polypeptide chains, which show decreased solubility are perfect substrates. Whenever, the solubility of the polypeptide chain becomes a problem, the use of tetraalkylammonium hydroxides in saponification becomes clearly the method of choice. This base is superior to other commonly used bases such as NaOH and LiOH. The ester **24** was nearly insoluble in either DMF or THF, however, the addition of tetrabutylammonium hydroxide (at 0 °C) results in fast dissolution of the solid and formation of a clear solution. There is a noticeable remarkable solubilizing effect from tetrabutylammonium hydroxide, which is probably responsible for the success of this reaction. The acid Boc-K(Z)LLLLK(Z)LL-OH (**25**) is obtained in 94% yield containing only about 0.8% of the undesired diastereomer (Scheme 7).

Boc-K(Z)LL-OMe **9** —[1. NaOH, acetone/H_2O; 2. H^+]→ Boc-K(Z)LL-OH **23**

—[HCl • H-LLK(Z)LL-OMe; HBTU/HOOBt, DIPEA/DMF, -4 °C]→ Boc-K(Z)LLLLK(Z)LL-OMe **24**

—[1. $(n\text{-Bu})_4NOH$, H_2O, THF, 0 °C; 2. H^+]→ Boc-K(Z)LLLLK(Z)LL-OH **25**

***Scheme** 7: Synthesis of the 1-8 acid intermediate*

We applied this hydrolysis procedure to many polypeptide esters, and for large-scale production of several polypeptide acids.[7] An example that shows the advantage of this procedure over other procedures is the hydrolysis of the simple nonpolar polypeptide ester Boc-LLLLLL-OMe. Using 40% aqueous $(n\text{-Bu})_4NOH$ (2 equiv.) in DMF at 0 °C, the reaction is complete in 2h giving an isolated yield of 85% and diastereomer ratio of 99.4 : 0.6. Compare to the reaction with aqueous NaOH (2 equiv.) in DMF at 25 °C; only 50% complete after 3 days with diastereomer ratio of 98.8 : 1.2. This and other examples (7) show clearly that this is a superior procedure.

Boc-LLLLLL-OMe —[1. $(n\text{-Bu})_4NOH$ (2 equiv), DMF/H_2O, 0 °C, 2h; 2. H^+]→ Boc-LLLLLL-OH

Next, we studied the coupling of the two intermediates, Boc-K(Z)LLLLK(Z)LL-OH (**25**) and HCl-H-LLK(Z)LLLLK(Z)LLLLK(Z)-OBn (**17a**). In accordance with our goal of minimum changes, we started with the previous conditions in THF/NMP/H_2O/$LiBF_4$ using DIC/HOOBt/DIPEA at 0 °C to rt. The mixture of the two compounds **25** and **17a** in THF/NMP in presence of $LiBF_4$ was homogenized for a few minutes followed by the addition of H_2O to give a slightly thick mixture that was easy to stir with a mechanical stirrer. After adding the coupling reagents, the reaction proceeded to form a fluid yellow solution in a few hours and was worked up after 24 h. The yield of product **26** was nearly quantitative and the area % purity was about 87%.

Boc-K(Z)LLLLK(Z)LL-OH + HCl.H-LLK(Z)LLLLK(Z)LLLLK(Z)-OBn
25 **17a**

DIC, HOOBt, DIPEA, LiBF4 / NMP/THF 20-25 °C, 96% → Boc-K(Z)LLLLK(Z)LLLLK(Z)LLLLK(Z)LLLLK(Z)-OBn
26

***Scheme 8**: The 8 + 13 coupling*

Based on the noticeable solubilizing effect of tetrabutylammonium hydroxide in the hydrolysis of insoluble esters, we selected tetrabutylammonium chloride as an alternative additive to solubilize the reactants. The results exceeded our expectations, the combination of the two intermediates, **25** and **17a** in THF/NMP/H_2O in the presence of n-Bu_4NCl (4 eq.), without homogenization, stirred at rt gave a clear solution within 1-2h. We could dissolve the intermediates in NMP alone, or a combination of NMP/THF or THF/H_2O using either the HCl salt (**17a**) or TFA salt (**17b**). The coupling reaction was carried out using DIC/HOOBT/DIPEA to give the product **26** in nearly quantitative yield and the area % analysis showed a purity of 92%.

Boc-K(Z)LLLLK(Z)LL-OH + HCl.H-LLK(Z)LLLLK(Z)LLLLK(Z)-OBn
25 **17a**

DIC, HOOBt, DIPEA / THF/H_2O, n-Bu_4NCl 20-25 °C 96% → Boc-K(Z)LLLLK(Z)LLLLK(Z)LLLLK(Z)LLLLK(Z)-OBn
26

Scheme 4: Modified 8 + 13 coupling

As in the first synthesis,, the final step was the removal of all protective groups under hydrogenation conditions in TFA/AcOH/H_2O solvent mixture with

Pd/C catalyst. Since the Boc-group is not removable by catalytic hydrogenation, the procedure was modified to dissolve the fully protected KL_4 in neat TFA prior to addition of water and AcOH and finally the catalyst. This initial TFA treatment removes the Boc-protective group. The catalytic hydrogenation was then performed to remove the other protective groups. The crude product from the convergent approach was consistently obtained in higher purity and consequently higher isolated yields. The final product KL_4 was purified by preparative HPLC chromatography and isolated as the acetate salt.

The advantages of using these two key intermediates in a convergent synthesis include:

- Control of epimerization, particularly at the 3, 8, and 13 residues to minimize the formation of the *D*-diastereomers at these positions.
- Easy in-process control of the two key intermediates.
- One critical coupling at end of synthesis leading to higher purity and lower risk.

Conclusion

The solution phase synthesis of the pulmonary surfactant 21-amino acid polypeptide, KL_4 was achieved in two different synthesis approaches. The first approach was a linear fragment synthesis that suffered from the low solubility of the larger fragments latter in the synthesis. A convergent approach was developed to minimize these effects and improve both yield and purity of the final product. An efficient procedure for the hydrolysis of large polypeptide esters with tetraalkylammonium hydroxides was developed and applied successfully to the synthesis of the key 8-amino acid polypeptide intermediate **25**.

References and Notes

1. Cochrane, C. G. and S. D. Revak, S. D. *Science*, **1991**, *254*, 566.

2. In these abbreviated representations, a free amino group is represented by H on the left side of the polypeptide chain, while a free carboxy group is represented by OH on the right side of the chain. The protective groups, Boc- and Z- as well as the benzyl (Bn) ester are drawn in the appropriate location on the chains. For example:

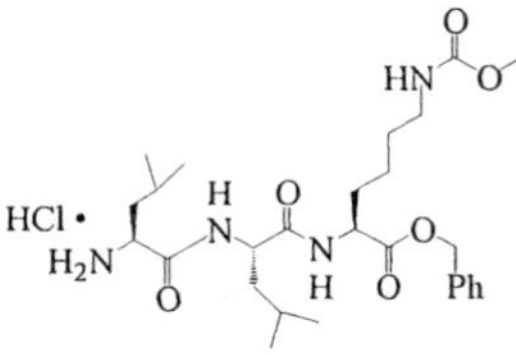	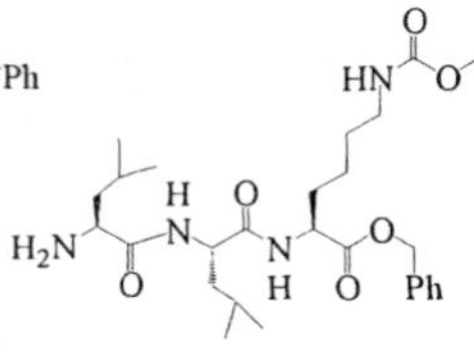	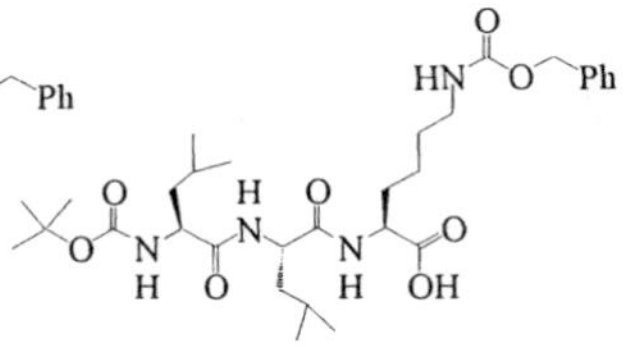
HCl • H-LLK(Z)-OBn	H-LLK(Z)-OBn	Boc-LLK(Z)-OH

Further abbreviation may be used for larger fragments, for example Boc-LLK(Z)LLLLK(Z)-OBn (**14**) may be described as Boc-14-21-OBn; the numbers 14-21 indicate the positions of the amino acids in the final chain.

3. The relative tendencies of activated residues to racemize during coupling of segments in dimethylformamide. N. L. Benoiton, Y. C. Lee, F. M. F. Chen *Chem. & Biol. of* Peptides, 12th Am. Peptide Symposium 496-498, **1992**.

4. Convergent Solid-Phase Peptide Synthesis. Lloyd-Williams, P.; Albericio, F. and Giralt, E. *Tetrahedron*, **1993**, *49*, 11065.

5. Reagents:

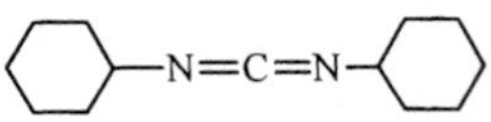	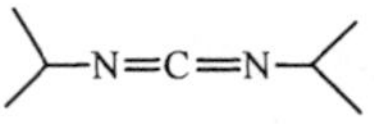	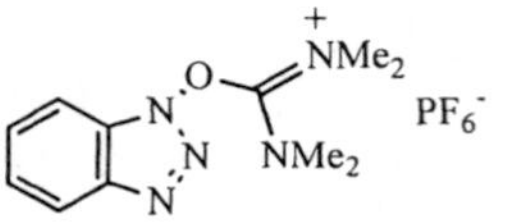
Dicyclohexyl carbodiimide (DCC)	Diisopropyl carbodiimide (DIC)	O-Benzotriazol-1yl-N,N,N',N'-tetramethyluronium hexafluorophosphate (HBTU)
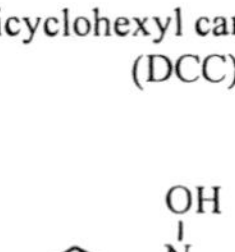	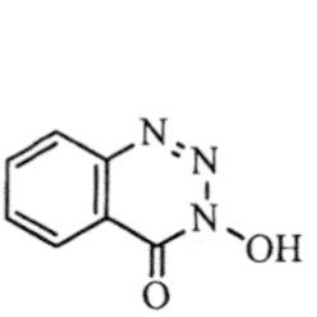	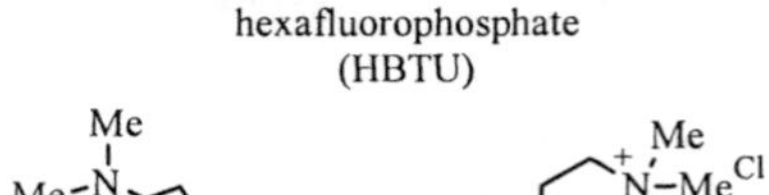
1-Hydroxybenzotriazole (HOBt)	3,4-Dihydro-3-hydroxy-4-oxo-1,2,3-benzotriazine (HOOBt)	1-(3-Dimethylaminopropyl)-3-ethyl carbodiimide • HCl (water soluble carbodiimide)

6. The diastereomers formed from epimerizations are labeled with a letter *D* denoting the *D*-configuration at the epimerized carbon and a number representing the position of the amino acid residue in the final KL_4 polypeptide chain, e.g., *D*-18 Boc-14-21-OBn refers to the *D*-epimer at the α-position of the leucine residue number 18 in the final KL-4 chain and is

represented as Boc-LLK(Z)L**L***LLK(Z)-OBn. In the structure, the asterisk (*) is on the right side of the letter representing the amino acid residue with the *D*-configuration.

7. Hydrolysis Of Polypeptide Esters with Tetrabutylammonium Hydroxide. Abdel-Magid, A. F.; Cohen, J. H.; Maryanoff, C. A.; Shah, R. D.; Villani F. J. and Zhang, F. *Tetrahedron Lett.* **1998**, *39*, 3391

Indexes

Author Index

Subject Index

B

C

D

E

K

L

M

N

O

P

Q

R

S

T

U

V

W

Z